HUMP PILOT

HUMP PILOT

Defying Death Flying the Himalayas During World War II

by Nedda R. Thomas

History Publishing Company
Palisades, New York

Copyright©2014 by Nedda R. Thomas

LCCN: 2014956725
ISBN: 978-1-940773-20-9 (softcover)
ISBN: 978-1-940773-20-9 (ebook)
SAN: 850-5942

Thomas, Nedda R.

 Hump pilot : defying death flying the Himalayas during World War II /
by Nedda R. Thomas. -- First edition. -- Palisades, New York : History
Publishing Company, [2014]

  pages ; cm.

  ISBN: 978-1-940773-20-9 (softcover) ; 978-1-940773-20-9 (ebook)
  Includes bibliographical references and index.
  Summary: Based on the true life exploits of a World War II pilot
flying the dangerous route over the Himalayas, the books brings to
light a little known facet of World War II. "Flying the Hump" was the
name given by American pilots to flying over the treacherous air currents of the Himalayas during World War II. It was an extremely dangerous but necessary route American pilots traveled to bring vital
materiel to Chinese troops in China, and American, and other Allied
forces in the Pacific. The materiel transported, critical to the Allied
war effort in the early days enabled the Allies to persist while the
industrial might of the United States was retooling.--Publisher.

  1. Thomas, Ned, 1923- 2. World War, 1939-1945--Aerial opera-
tions, American. 3. World War, 1939-1945-- Campaigns-- Himalaya
Mountains. 4. World War, 1939-1945--Campaigns-- China. 5. World
War, 1939-1945--Campaigns--Burma. 6. World War, 1939-1945--
Campaigns--India. 7. Himalaya Mountains-- History, Military. 8.
China--History, Military. 9. United States. Army Air Forces--
Biography. 10. Air pilots, Military--United States--Biography. 11.
Airlift, Military--United States--History. I. Title.

D790.2 .T46 2014
940.54/4973--dc23   1411

Published in the United States of America by
History Publishing Company, LLC
Palisades, New York

Printed in the United States on acid-free paper

First Edition

To my son Ned Davis
His grandfather's namesake

TABLE OF CONTENTS

FOREWORD...ix

PROLOGUE—OLD MEN DON'T FLY THE HUMP............xv

CHAPTER 1—"A LOT OF HIGH, HARD ROCKS"..............1

CHAPTER 2—"NOOKS"..28

CHAPTER 3—"PRIVATE PROPERTY: NED THOMAS".... 49

CHAPTER 4—FLYING THE FIREBALL.............................71

CHAPTER 5—HOUSE OF SNOW....................................97

CHAPTER 6—"AN EPIC OF THE WAR"........................ 129

CHAPTER 7—"IF THE PLANES WERE READY,
                        WE FLEW"... 157

EPILOGUE—PEEL BACK THE SKY.............................. 177

NOTES..185

BIBLIOGRAPHY..187

INDEX..189

ACKNOWLEDGMENTS

The author gratefully acknowledges many who supported the mission. Eda Kenny of the WWII Flight Training Museum at Douglas, Georgia, who shares a trove of insight about vanished aspects of air cadet training; also Danny Bradley and Denise Mortoff, of the same, for their help and interest. David Selby, photographer extraordinaire and supportive military enthusiast; Leigh Kitcher for technological support beyond my capabilities, both of Historic Vienna, Virginia, Inc. Alice Nodine, Editor, Elan Magazine, for encouragement from the start and always. Historian Naomi Zeavin, for her helpful early reading. Author Lou Thole for his generosity and enthusiasm. Robert F. Dorr, military "superfortress" who shares the writing universe. Evelyn Gardett, my sweet last-minute proofreader. Adriana and John Hall, loving friends through smooth flying and rough. Lt. Colonel Elizabeth Ortiz, Director, National Media Engagement Office for the Secretary of the Air Force, Office of Public Affairs: also Deputy Director David J. Wilson, their help in locating relevant photography. Stephen Roberson, webmaster/solver of the unsolvable, for all things computer. My publisher Don Bracken, and the History Publishing Company, for believing in this story. Military historian Walter J. Boyne, former Director of the National Air and Space Museum, for his great and generous heart, and knowledgeable Foreword. Finally my father, Ned Thomas, who sat down with me amid the plaster and debris of a house rehab, and taught me to fly the Hump.

FOREWORD

by Walter J. Boyne

TIME, THAT INEVITABLE, IMPLACABLE MONSTER, HAS TAKEN ITS toll of what is now commonly called "the Greatest Generation." Day by day, the men and women who fought and endured one of history's greatest tragedies are quietly slipping away. Fortunately, there remain some who recall the events that so troubled the world from 1939 to 1945, and who can tell how their personal experiences affected their lives, their war, and indeed, our history.

Such a man is Ned Thomas, who gives us an intensely personal look at a life that included something he had only dreamed of —becoming a pilot—and then thrust him into the most primitive, most demanding , and most overlooked of all the theaters of war. Fate steered Thomas to become a pilot on what became too familiarly known as "the Hump," the great airlift that managed to keep China in the war against Japan for three vital years.

The term "too familiarly known" is used advisedly. There have been some books, films and television documentaries on the men and planes who flew across the Himalayas from India to China—just enough to reduce the phrase "the Hump" to familiar and often dismissive jargon. This book puts the massive, highly successful operation into a context that swirls with conflicting national motives, narcissistic leaders who made enormously important decisions based on their own selfish desires, and often, bribery on a grotesque scale. There were tremendous aerodynamic challenges. These forced pilots to operate routinely beyond the capabilities of their equipment, with all too often fatal results. The route over the Himalayas was not called the "Aluminum Trail" in humor, or with sarcasm. It was an appropriate designation for a highway clawed into mountains and jungle, its markers the 700 destroyed aircraft and the many dead crews.

The effort started small, from hastily packed bundles stuffed into bomb bays and launched from rock-strewn runways in India over the mountains to Kunming, China. It grew slowly at first and then ever more swiftly into a fantastic aerial conveyor belt delivering every weapon, skill and service. Ironically, this great supply effort overcame the confused battle of egos, national honor, and rampant graft. It succeeded despite being conducted over the most hostile combination of geography and weather encountered anywhere during World War II.

Perhaps the most important, and certainly the most winning, aspect of this book is that it proves conclusively that the operation was massively successful because it depended first, last, and always upon its personnel at the lowest operational levels. It worked because of people like Thomas, who accepted the responsibilities laid upon them and, ignoring the practical impossibility of operating at all, went on to achieve success on an unprecedented scale.

More than seven decades have passed since President Franklin Delano Roosevelt realized that the brilliant invasion and capture of Burma in 1942, by Japan, had eliminated China's last overland supply link, the famous "Burma Road." He decided that at whatever the cost, China had to be kept in the war by means of an aerial resupply system. The politics were ugly. Nationalist China's leader, General Chiang Kai-shek, was more concerned with fighting his Chinese Communist opponent—

Mao Tse Tung—than he was fighting the Japanese. Chiang Kai-shek was detested by the local American commander, General "Vinegar Joe" Stilwell. The Soviet Union, preoccupied with its Nazi invaders, wanted Japan to be more engaged by China, so that it could move forces from East to West. If China fell, Japan might well attack Russia and aid the Nazis.

And then, to make Roosevelt's decision more difficult to enforce, there was a complete lack of suitable bases, transport, aircraft, pilots, mechanics, and all of the millions of tons of equipment that are implicit in establishing a new theater of war.

Of all the difficulties, the combinations of weather, terrain, and inadequate aircraft exemplified the worst. The early bases were primitive in the extreme, lacking adequate runways, maintenance capability, living quarters and sufficient fuel storage. The planes were simply those available—Douglas C-47s, a few Curtiss C-46s, and the modified Consolidated B-24 lash-ups, the C-87 and the C-109. These planes were routinely overloaded and dispatched on missions over the Himalayas that ended all too often in disaster. Any malfunction—the loss of an engine, the failure of a supercharger to operate when required—became a death sentence over terrain in which crash landings were impossible. Many crews were forced to use their parachutes, descending into a hostile environment where survival was not problematic—it was usually a vain hope.

Hump Pilot succeeds as the author interweaves one man's story into the context of this "forgotten war." Readers see how his life, and doubtless those of other pilots, is challenged, changed, and enlarged. Military and political players are portrayed to reflect up-to-date research. Cockpit scenes, relevant aircraft and navigation, impart a sense of what it felt like to battle the brutal conditions of the top of the world. CBI history is as convoluted as any there is, but as a human story, the book stays simple and straightforward, and brings the focus back to Thomas as he experiences the unfolding events of World War II in Asia.

Author's Note

In discussing aircraft, navigation, and flight procedures of the 1940s, my aim is to convey essentials without excessive complexity, the goal being always to bring readers into the cockpit so they can experience the action from the pilot's seat. Any errors would be entirely my own. Dialog appropriate to the character's memory and reality has been "fleshed in" from time to time, to bring the action more fully to life, but all flights described, without exception, involve actual missions flown. Likewise, the names of commanders and world figures are real; however, for copilots or crew, pseudonyms are used.

OLD MEN DON'T FLY THE HUMP

Central China: November, 1945

NO NEED TO STARE AT HIS GAUGES. THEY SPELLED OUT THE whole grim tale at a glance. Fuel and auxiliary tanks on both wings officially empty, as of now.

Beyond the solarium windows of the cockpit, no landing spot snared his searching eye. Just the same shuffle of muddy villages dozing inside their walls of weathered rock. The eternal checkerboard of rice paddies reflecting the autumn sunlight. A serpentine skein of narrow roads to tie one tumbledown hamlet to the next, and twine an interminable, wavering thread through a million square miles of war-ravaged interior China.

High over this bucolic picture book, American pilot Ned Thomas had a big airplane and a big problem. Pick out someplace to land, or get seriously ready to ditch.

The flight had stretched uneventfully for several hours since takeoff from Hankow. A clean sky shined at his back, broad daylight ahead. The winds stood calm. Nothing out of the ordinary to brace for. Not that you tend to count on the ordinary in this line of work. He'd figured to wrap up a seventeen-hundred-mile mission to Kunming, set down his C-46, a winged workhorse if ever one was, and take on another detachment of Chinese troops. Sixty tired, scowling men, part of the great fighting force of their

Generalissimo, Chiang Kai-shek, straggling aboard with their gunnery and cook pots, cabbages and equipment, rice bags, contraband, and chop sticks.

He'd report to base-ops, swing into the mess for a bite, leave the load masters to distribute any cargo, then add a few special adjustments of his own before taking off back to Hankow, where the soldiers would deploy north for another strike at somebody or other. Maybe root out a few more Japs from their hideouts. Maybe beat the Reds farther into the gray countryside before everyone hunkered down for a mean winter.

The best laid plans.

World War II had ended in Asia, in name anyway, with an Armistice in August. Not that peace was evident in this weary, conflict-wracked chunk of the globe. It was hard enough just to distinguish victor from vanquished, if it weren't that the defeated Japanese lived so much better. He saw them idling at captured bases, the same men, who if he'd met them over the Hump, would've been coming at his windshield in a Zero. He watched them strut along in pressed uniforms, boots polished, swords dangling, hale and cocky among the populace who'd suffered a generation of terror and abuse under their occupation.

As far as he could see, the only change wrought by the armistice was that hostilities no longer embroiled this huge nuisance presence of Japanese throwbacks. Suddenly it was Chinese against Chinese. Or any American unlucky enough to get caught in their cross hairs. Or to crash a plane somewhere beyond amenities like help and rescuers. You think you're ready to pack in a good flight, and you never know.

By habit he'd been scanning his controls, alert in the familiar roar of the 2000 horsepower engines, the machinery odor of ozone and oil, the vaguely scorched electrical smell emanating from the heat of the radio tubes. Keeping Argus eyes on his dials, in sharp relief against the panel under the wide cockpit windows, he watched his oil pressure, temperature gauges, altitude, air speed, things a seasoned flyer checks by second nature.

Experience taught that if the weather turned clear, rarity though that was, still you look out and appraise the skies with the same watchful eye, heedful of cloud formations, particularly the thunder heads of the cumulonimbus that loom up like anvils, which you fly around if you can, because the monsters can tear a plane apart. You look down the lower window at foot level and survey an alien world, presently consisting of rutted paths and the occasional hand-pulled cart. You look up again because for sure, you don't want to run into another airplane strewing its tracks across this back of beyond. You keep scouring your environment, the envelope that changes as you fly. You can hit snow and sleet, wind shears and turbulence. On instruments you'd be checking the artificial horizon, too, but visibility today is uncommonly fine.

"Say, Ned." Over the decibels his copilot's voice drifts from the right seat. Some guy sent out from a stateside job at an air-field in Indiana. Or is it Iowa, or Idaho? Pleasant, maybe a bit green.

"Betty Grable, there's a dame you gotta like. Smart as a whip, and pretty! Or if she's not your dish, how about Hedy Lamarr? Why, she must be the most beautiful woman in the world. Soon as we get home, buddy, I'm finding me a girl who'll knock your eyes out."

"You're looking at the decrepit ones, pal. Grable's twenty-nine if she's a day."

Shifting a boot around his left rudder pedal, Ned flexes shoulders locked in tension after hours of flight. He checks the mani-fold gauges, something he could direct to his copilot except his is talking. Not that you slack off. It's still almost a novelty not to need the oxygen mask he'd worn above the high snow-shrouded Hump, the Himalayas, in the turbulence, on instruments, the air a thick unseeable ocean dark as ink, where inside an ice-rimed cockpit, he'd waged a bullfight at the controls while frozen shards and ice chunks flung off the props and crashed in his face, and lightning danced across the wings. The daily brush with death. The good men he knew who went down.

"Tell you what, Ned, I wouldn't care if she's thirty. Make that a hair shy of thirty. I like a woman with some maturity, specially one with a pretty all-American smile and great legs."

"Check 'em out, lieutenant. Warn 'em, too, so they don't swoon when they see you."

"That an order, Capt'n? I got a reputation I don't want to muddy up."

"Well, use your discretion. An officer and a gentleman. I chose my girl before I pulled out of the States. Feel free to take your pick from who's left." No need to add that his Neva Rae has a face to put all those movie stars in the shade.

"Hey, Ned."

"Yep?"

"While back, you said you always make a few extra arrangements before you'll fly old Chiang's boys. I thought the sergeants took care of that."

"Not quite."

"What's left?"

"I confiscate their weapons. Stash 'em in the lower cargo."

"They wouldn't start firing inside an airplane."

"It's been known to happen."

"And they obey? Those beady-eyed killers hand everything over to you, just like that?"

"I'm not taking off till they do. They're not putting a hole in my plane. Or in me."

"What kind of weapons would they pull out on a plane?"

"The exploding kind. Guns, grenades. Bayonets, too."

"You aren't serious."

"All a guy's got to do is pull the pin."

Static scratches through the radio. He reaches overhead to tease the volume. The abrasive garble could've come from a far-away planet except for the anaemic voice stringing along.

"Hankow tower calling flight One-Eight-Two-Six-Dog, over."

Dog is his squadron call sign.

"Roger, Hankow, this is 1826 Dog. Over."

"Mission scrubbed, one-eight-two-six. Return to base."

"Roger, Hankow. Over."

Funny, they've come this far. But orders are orders.

"What d'ya figure's going on, Ned?"

"Beats me, but we're turning around. Passed midpoint some while ago."

Giving his copilot a thumb-up to hit the throttles, he reconnoiters. To get back to Hankow they need to refuel, something he expected to look after at their terminus in Kunming.

Now they've attained two-thirds of the way on a run where fuel reigns a couple notches above the Holy Grail. Enough to get where you're going, no farther. Because high-octane petrol, that's what feeds a war. He and other pilots flew the incendiary materiel over jungles, rivers, the Top of the World. Operated on margins so thin he tried not to think about them. Risked their lives in burdened planes packed with leaky, outsize drums. Climbed to reckless, insane altitudes, knowing they could go up in a fireball with a flash of lightning, a spark from an engine, a pop of enemy fire. They steered these airborne gas pools through hostile airspace in the worst weather on earth, as poison fumes curled smoke-thick inside. Flying coffins. The death count and lost planes told the story.

Today he's just flying across a pretty sky on empty tanks.

"Say, Ned, want to pull up higher? Get the big picture."

"No, hold her steady."

Don't tax the lines. Can't spare the juice. How about Chihkiang? He knows the way. Deployed and lived there in a tent coming out of the Hump. Not much off course, might add an hour extra, all told. He'll divert, refuel, then resume the northward leg to Hankow. Adjusting his earphones, he reaches for the overhead console to tune the radio to frequency.

"Chihkiang tower. This is 1826 Dog, over."

No reply ensues, only a sandy crackle. Not that a scratchy signal's unusual. Let the winds hold, soon he'll taxi in, fill his tanks, and resume the upward jag to Hankow. Arms and legs in command of his controls, throttle, rudder pedals, he starts banking the ship, rocking in the current of a giant cradle, steady in the roar as he nudges her out in increments.

"Chihkiang tower, this is 1826 Dog, over. Come in, Chihkiang."

Shouldn't be long, they're flying smooth. Being behind the battlefront, he's not looking for ground fire, that's the good news. No fuel to go acrobating around a spray of flak. He tweaks the frequency, tries again. Spears a glare at the clock on the panel, the hands advancing passively round the dial. Ten minutes. Fifteen. Tick-tick. The sun shines bright.

"This is Chihkiang tower. Over." About time.

"Chihkiang, Army Air Force Dog 1826. Request clearance to land for fuel, over."

"Chihkiang, to 1826. No aviation fuel available at base. Over."

"Roger, Chihkiang." He grips the yoke hard. Vacation's over.

Having lost vital fuel on a wild-goose switchback to Chihkiang, too far to bend rearward to Kunming, his sole present option is to fly back to Hankow on what's left. If he can.

Hunkering into the mass of the aircraft, he guides the plane into a one-eighty. The engines emit a throaty yawl, the '46 bends on its behemoth track and pulls out.

Gotta reduce power settings to maximize fuel. Can't climb and squander dwindling reserves. Do the critical adjustments to throttle and props, yank the airspeed to one-forty, and hold it as low as he dares without hitting stalling speed. Cut the RPMs and baby the ship. Under a Hump load, popular lore would've had him handing over the controls to his copilot about now, getting back to the cargo door, a rope knotted round his midsection, killer winds tearing his skin off, kicking gas drums, anything but humans, out over the jagged peaks to lighten the plane. As it is, he'll just have to squeeze the last driblet from every tank and wring 'em dry.

Laying hold the wheel-over-wheel tank selector, he begins the dicey process. Fly the first to dead empty, switch to another. Burn each by turn down to the fumes. Do-or-die. On fuel starvation your engines will stall, so you don't blink till you transfer to the next tank. Keep doing it right, and you'll hear that barbaric cough as they fire again, that pulse of life.

If he's a sure hand with anything, it's a stall.

Turning the dial, he weighs which tank on each wing likely holds the most. Can't let one wing outweigh the other. A stricken plane, you gotta keep it balanced. Got to play every card right, to home base or the hereafter, whichever comes first.

His copilot yawns. Maybe nodded off. "You're gettin' kinda quiet over there, Ned."

Tensed forward, his parachute lumped in the space beneath him in the seat —an item he may need if they run dry before he can land —he pushes the plane till he empties the first tank. How long did it take? Never mind, go to the next. Pilots nowadays, they can regulate their fuel lines from the cockpit, thanks be to innovation. They're cruising so steady his eyes could pick out a shook of bamboo. All he wants is some level ground.

Now here's his copilot rousting from nap time.

"A few more flights, I'll be getting my rating. Be a first pilot like you, Ned, sitting over there in the left seat, running the show"

"Not if I don't land this thing."

"Hey, you're at the controls, who's worried?" The man shifts in his seat. "Look, I'm twenty-five and you might think that's young. But I catch on fast. Maybe you'll be the one to check me through. Say, how old are you anyway?"

"Turned twenty-two a few weeks back." Every hour of which would be flashing before his eyes if it weren't for needing to nurse the blessed tanks. His companion gives a low whistle.

"You gotta be kidding, old man. Hardly twenty-two, and you came out of the Hump?"

"Old men don't fly the Hump. We were all young."

"Well, they must've snatched you outta your mama's arms."

Thunk. Another tank shuts down. Quick, rotate the selector. You hear a thump from the plane like you hit a pothole in the sky, hold your breath for the new spark, how sweet the sound.

The lore of a belly landing starts to reel intrusively through his head. In a rice paddy? Hey, paddies are level. Got some dikes around to hold in the water, but those aren't so high. A big splash into one might beat a rock-hard crash any day. Or if he can get farther along, well, there's the call to the tower to clear the runway for a long, metal-screeching slide, then. . . .

No, don't think about it.

The tanks dry off in orderly sequence. The situation passes from attention-getting to dire. Fuel depletes and everything cogitable races through the backrooms of his brain. Get close and he might go for a glide before the plane falls under its lost momentum. Nope, no good, he needs altitude and speed to glide, a lot more, this plane's heavy. He doesn't have altitude, doesn't have speed, doesn't dare throttle up for a climb and diminish life-giving fuel. If worse comes to worst, he'd rather set the ship down someplace soft, not on some farmer's pigtail either. Just go down like a meteor hurtling in, and feel that outsized spatter as you connect with earth. A jolt to shred a plane and strew the pieces across a field. They might live to tell it. Men have.

A few.

Twenty minutes more squeezing the tanks, and he's straddling life-and-death for fuel. The place between needing to move faster to save time, and slow down to conserve.

Come on, Dumbo. Wing it like a bird.

At least he's not battling headwinds, a present point in their favor. But reaching the last tank, they're maybe eighteen minutes out of Hankow, and it was the lowest on either wing when he switched into the sucker. This flight has outstretched every fuel expectation. They're in real trouble. He's going to need speed now to make it all the way. Speed without increasing power. Which means setting up a slow descent ahead of schedule so he can gather velocity coming down. In fifteen minutes, he'll radio for a direct approach to avoid circling in the stack.

If he's got fifteen minutes.

Don't lower the gears yet. Can't add drag on the plane. Can't monkey around with other aircraft in the pattern either. Just labor in, get clearance to dive under them, and set her down.

Fifteen minutes dwindle to twelve. Ten. Five.

"This is Army Air Force 1826 Dog, calling Hankow tower. Over."

"Hankow tower. Come in, 1826. Over."

"Calling for landing instructions. Low fuel. Request direct approach to runway."

"Roger, 1826. Call on final. Over."

Four. Two. He's wrung it from both wings to the vapor. They're on final. Radio to tower. Runway nose-ahead. Drop the wheels.

The red gear lights on the panel flash to even green. See to the flaps, get a bit of lift, reduce stalling speed, but not too much. Time to set this bird on the ground.

They roar in like a rocket. The gears reach down for the hit. He pulls back on the throttles as they thunder across the fence and connect with the far end of the field. Touchdown.

In a wailing blur he taxies to the ramp, where the plane suddenly lets loose with a tremor end to end. The engines on both wings sputter and die conjointly, like Siamese twins. Like a deflated carnival dragon. Fuel starvation.

His companion stands, stretches, and saunters from the cabin.

By habit, Ned reaches to reset his throttles. Closes the mixture switch for the nonexistent fuel, shuts the magnetos.

A flight he never forgets. "We'd been in a critical situation. If my copilot ever got in the loop, and I'm not sure about that, he wasn't sweating like I was. I didn't know if we'd make it. Might have had to take the plane down on the way, but it would've been a mess. I lingered a few moments in the cockpit to reconcile my thoughts."

Opening his side window, he breathes in, steps from his seat, and swings down the darkened cabin to the door, where the jeep idles to drive them to the dilapidated ruin that serves for Base Operations. He'll close out his flight plan, then head for the bombed-out hanger where pilots bunk. No heat in the place, no roof. Maybe there'll be a letter waiting from home.

It's on the jeep that he experiences a graphic flash of hindsight.

"I should've refused to turn back and kept on to Kunming. I was the captain of the plane. I could've said I didn't have the fuel to do it. In fact, I probably was derelict not to do it."

For a pilot who'd been forged in the unforgiving crucible of the Hump, the lessons died hard. And that's where his war had begun. Seasoned by the highest altitudes on earth, by tough, tortuous flying under unimaginable conditions and unending danger, he'd got through it day by day, night on night. Never bucked an order. Not once. Scared? Not on your life.

Apprehensive maybe. On occasion. Flying planes shorted on fuel and routinely overloaded. The insane, defiant mantra echoing: There will be no weather over the Hump.

Worst joke of the war.

Signing off, he's conscious of a weariness. Can't let it bend you. Ned Thomas, first in his class to solo, first to turn his goggles forward, a military flight instructor. Trained on the big planes. Lived through the Hump. If they'd gone down, he and he alone, would have been responsible. Crazy, how much can happen in one day before a man even gets his lunch.

Hearing a voice at his back, he straightens. Here comes his CO, John Langford, trim frame erect, familiar chin jutting out, mouth set in a fold below a tidy, Don Ameche mustache. Rattling a sheet of paper in the air.

"Ned Thomas."

"Major Langford."

"Is it idiots all over the place? Every last one of them. The Orient Project, they say. Gotta save Chiang, they say. Every man's life on the line, they say. And what for? So you can come to this?" He shakes the paper in Ned's stunned face.

All right. Today he was an idiot. "Yes, Sir."

"Blast them all to kingdom come. Won't Washington ever see past its nose? Always switching gears on you. Ol' Willie can't sell 'em on one plan, or fix another."

That would be their commander, General William Tunner. World's foremost detail guy. Has he got his big fist in today's close call? Unbelievable. Bad news sure travels fast.

"Did they understand what we were up against in India? Not a chance. Do they want to hear how much we still need you guys in China? Well, you can forget that, too. Just like they never had a clue about flying the Hump, not from who shot Lizzie.

"You could do a lot here, Ned, given the chance. You gotta know it. Others, too, but you've been one of my best. Now Washington's on the teletype making another mess on top of a mess, and we gotta play by their rules."

One insane day. Yeah, he took a risk. But somebody sent out a wire?

"So locate your gear, Lieutenant."

Sounds like his flying days are over and done. Is that him gulping? "Yes, sir."

"And be ready to pull out for good."

"Yes, Major." He never gulps. Hard to think, his career smashing to an end this way.

"Hankow base is shutting down. We've got orders to move all

aircraft to Shanghai, and roll up the runway. You know the story, pilot. Catch yourself a couple winks and get set to fly."

CHAPTER 1

"A LOT OF HIGH, HARD ROCKS"

Wake County, North Carolina: November, 1942.

IN THE GATHERING LIGHT OF AN AUTUMN DAYBREAK, TWO TALL young men threaded purposeful strides across the campus of old Wake Forest College. Gaining dual-lane Highway 1, they began to hitchhike toward the State capital at Raleigh, some fifteen miles south. Neither broke the silence of the early hour to comment on the faded fields bare of their harvests of cotton, corn and tobacco, the weathered tobacco barns scattered among the low-lying curves of the land, the whitewashed farmhouses embraced roundabout with country porches. These were familiar scenes. Their attention was focused on what loomed ahead, an officer-candidacy exam, the first hurdle they had to clear to become cadets and, so they hoped, to qualify to train as wartime pilots.

Fraternity brothers, roommates, best friends in the opening semester of their junior year, the one with the serious face and suntan was twenty-one year old Deming Ward, who hailed from Durham. His companion, ready with a smile for sympathetic drivers in an era of rationing and scarce cars, was Ned Thomas. Having turned nineteen, Ned, too, was tall, a shade over six feet, with strong regular features and a lively up-quirk at a corner of

1

his mouth. He might have passed for a bit older, having finished high school in his home town of Roxboro in three years and begun college ahead of his age. Both now had completed the four semesters required by the U.S. Armed Services to apply for officer training. Provided they did well on the test. The draft was breathing down every man's neck, and one way or another, they took them all young these days.

The business that had the pair thumbing to Raleigh, dovetailed in uncanny precision with events unfolding between their nation's capital in Washington, DC, and China, a remote land enwrapped in such mystery that most Americans could at best have named it as a vast faraway country beset by destitution, conflict, and starvation. Yet as Deming and Ned advanced along the highway, their President and his chiefs were busy hammering together a venturesome project to save this unpromising land, with its incomprehensible mishmash of war and misrule. Their goal: Establish an airlift over the Himalaya Mountains to the blockaded Nationalist Chinese and their leader Chiang Kai-shek, America's sole ally in Asia still fighting the Japanese.

From the time of Japan's strike on Pearl Harbor that past December 7, 1941. if not before, evenings found the fellows gathered at the radio set in the fraternity living room to hear news of the aggressions foisted upon Europe by Nazi Germany and Fascist Italy, and in the Pacific by Imperial Japan. No mistake, the U.S. faced deadly and belligerent foes on two fronts. Their mothers and fathers had listened in disquietude and shock when in 1931, the Japanese seized Manchuria and began their path of conquest. Unstoppable and remorseless, in a mere six years, that is by 1937, they blockaded the entire coast of China. In addition, they'd seized control of most of the central country, including Nanking and Hankow to the east, the Canton region in the south, and Peking to the north. All in all, half of China's people and 95 percent of her nascent industry were under the heel of Imperial Japan.

In the first six months of the present year of 1942, Japan mowed down the last American outposts at Guam and Wake Island. Next in line came the Philippines, where on April 9, U.S. troops surrendered and began their horrific death march from Bataan. Resistance broke again when the aggressor seized and occupied British Malaya and Dutch-ruled Indonesia. In May, they barely were stopped at Australia's doorstep, in the hard-fought Battle of the Coral Sea. From the central Pacific, U.S. servicemen were pushing toward the Japanese in a series of island encounters including the one General MacArthur led at Guadalcanal in the Solomons, that August.

Then Burma fell. Without this neighbor to the south, China's Nationalists lost access to the Burma Road, with its 700 miles of graveled and maniacal switchbacks that coiled like a hummocky, extraneous bedspring from Lashio, through the jungles and into the foothills of the lower Himalayas to Kunming, Chanyi, Chengtu, Nanning, and other inland ports. Completed in 1938, the route had been Chiang Kai-shek's last remaining route of supply overland. Retreating to his mountain capital at Chungking in western Yunnan, urgently he appealed across the sea to the Arsenal of Democracy. His forces were hungry, isolated, and beleaguered, his prospects of remaining in the war hanging by a thin and fraying thread.

Recent in Ned's and Deming's minds that day —for the event took place a couple of weeks before the exam—were a series of simultaneous surprise attacks that Japanese fighters launched on both sides of the nascent Himalaya air route from China, across the southerly range into the Assam Valley of India. Against staggering odds, and almost at Kunming, they were turned back by General Claire Chennault's legendary air fighter group, the Flying Tigers. Writing for Life Magazine about the Tigers' brash American commander, Clare Booth Luce describes the fifty-one-year-old Texan, with his consuming passion to beat the Japanese ". . . this wrinkled, scar-faced, half-deaf, ex-barnstorming pilot, the one genius that war on the Asiatic mainland had produced."[1]

Allied bases on the India side, Dinjan and Chabua among them, sustained bombing that destroyed runways and disabled aircraft before pilots could get them off the ground. The October 25 assault might have wreaked sufficient destruction there and then to demolish all hopes for the trans-Himalayan air-lift, had the Japanese not flown on instead to bomb the Indian docks at Calcutta and Chittagong.

Varied considerations then, had President Roosevelt studying Generalissimo Chiang's dilemma. He genuinely empathized with the Chinese people. Recognizing America's own critical situation in the Pacific, he could not ignore that in excess of half Japan's armed forces—2 million troops—were engaged within China. These factors made the country the foremost strategic square on the chessboard of the Orient. Roosevelt knew that to sweep the Japanese from America's western gateway, he'd have to divide and weaken them, a task he hoped the Chinese Nationalists' ongoing war effort would accomplish. To support them, the U.S. would assign two air groups to fly from India, high over the inclement and inhospitable Himalayan Mountains, and down into China. These outfits constituted the Air Transport Command, or ATC, and what at the time was called the First Ferrying Group of the Tenth Air Force.

Several flight patterns were proposed. Few proved feasible. With Japan in possession of the vital allied airfield at Myitkyina, Burma, pilots had no choice but to contend with a northerly course, higher mountains, and worse conditions than to the south. The track involved distances of a thousand or more miles round trip, with altitudes upward of 25,000ft in non-pressurized depression era planes built for calmer climes, and whose engines did not operate capably under such conditions—all in range of Japanese bases and their Zeros.

The route was homicide. Men had to fly the stormy Himalayas on instruments alone, in weather encountered nowhere else on earth. Every mile of the way was a life-and-death struggle with geography, atmospheric challenges, poor-to-nonexistent radio signals, lack of bases and crews, maintenance and supply leaks, aircraft shortages, the hopelessness of rescuing those who were downed—the appalling list went on. U.S. Lend-Lease designated for the Nationalists couldn't traverse China's barricaded borders except by air, and the ATC and the Tenth were flying with no back-up other than Chenault's overstretched Tigers, whom they supported and supplied.

Determined not to let the airlift succumb to its birth pangs, Roosevelt firmed up his commitment to Chiang Kai-shek to get American supplies across the mountains. Perhaps nothing reflected the President's faith in flight so much as this audacious and extraordinary plan in 1942, to implement a massive air operation over the most colossal mountains on the face of the earth. Owing in no small part to the decision, the commander of the Army Air Forces, General Henry "Hap" Arnold, later would call FDR the best friend the Air Force ever had.

In North Carolina meanwhile, the pair of friends from Wake Forest passed their exams with flying colors. Person County's Draft Board congratulated Ned for scoring exceptionally high. Deming, he assumed, did better, being in his estimation the "brighter" of them. New Year brought wires from the War Department containing reporting instructions to Miami, Florida, one among many basic-level facilities cobbled together around the nation to move men into active service, and the sole military posting the two would have in common until the Hump.

The die was cast. A telegram set the story in motion—a brave man, a harsh place, a global conflict that kept the elements in a steady march. Like millions of others launched into service, from California to Texas, the Midwest to the Eastern Seaboard, each was different by the unique, untold path made of objects in his common life. But all shared the bond of their willingness to sacrifice and give their best to defend the land they loved.

What of the place where Ned would be sent? Few could have told him much in that year of 1942. War correspondent Eric Sevareid, covering the China-India-Burma Theater, later wrote of the official interdiction barring early Hump pilots from letting their families know where they were or what they did, leaving loved ones to believe they'd been sent out on routine transport assignments. "It sounds like a cinch," Sevareid reports when security permitted, "but I have heard our fighter pilots and bomber crews out there talk about the Hump fliers, and I know the deep respect in which they are held. I remember Lieutenant Tommy Harmon, who fights in a P-38, saying: 'I would rather fly a fighter against the Japs three times a day than fly a transport over the Hump once.' They are carrying gas, bombs and ammunition for General Chennault . . . and jeeps, guns, medicine and a thousand other items to keep China's flickering resistance alive.

"The job takes a particular type of lad. His navigation must be excellent; for most of the run he must maintain radio silence, and it is easy to get lost over that rugged terrain, especially among the towering ice-covered peaks." [2]

Could Ned have been given a supernal preview of the future bearing down, or of the other side of the world where he was headed, he'd have seen himself moving with unimaginable impetus. From cadet life, pilot training, wings and commission (even unexpected romance) into a grueling wartime role, every step would fuse steadily into the next. After he reached his alien and hostile destination, he'd be flagged through with the same momentum, into the left seat as a first pilot after just two check flights over the Top of the World, testimony that he was recognized again and again as a very capable flyer. Correspondent Sevareid's words speak with prescience about men like him, for often the fighters attracted the "glamor," while outfits like the ATC got belittled. "Allergic To Combat," some said. But no insults got flung at flyers in the Hump. The route was an airborne hell. Even passengers who crossed the range, perhaps one time, would come home to boast of having "flown the Hump." Other pilots thanked their Maker not to be among those sent out to the deadly range, and they held in highest respect any man who was.

So challenging was the Hump that a supportive process evolved from the earliest days, whereby pilots with experience shared the ropes and taught the newcomers, who first flew right-seat with them. Some pilots never checked out or got put in charge of their planes, a crushing letdown. "No one wanted to be relegated to the right seat forever," in Ned's words. His own check pilot was a non-commissioned flight officer (a rank discontinued after the war), a senior man from his view, stocky with an old-time handlebar mustache, and like Claire Chennault, a Texan. The non-commissioned grade caused Ned some thoughtful reflection. "I always thought the rank of flight officer was unfair. They did the same work and should've had the same rank." Actually, the designation tied in large part to severe pilot shortages. As for mustaches, stubbles, and a less-than-parade attitude, commanders on the scene shrugged and let pilots be. Little sense it made to demand pressed shirts and a clean shave on men who flew from the monsoon, bug-infested tropics of humid India, into glacial elements aloft, down into cooler China, then the whole thing in reverse, in a single day or night.

Would any flyer ever forget his first view from the cockpit of the Roof of the World? Not likely, though seldom did the scenery open on an initial flight owing to the abominable high-altitude storms. Just to see this surreally beautiful terrain—these giants in the earth with their jagged storm-shrouded peaks and river-swollen gorges, their vast heart-stopping reality—beggared words. Pilots assaulted the range in propeller aircraft stretched beyond the limits of altitude capability. They flew day and night, in storms and under conditions unprecedented for flight. They hit air masses and fierce winds that flung large planes around like feathers in the sky. They felt the floor drop away in jet streams that threatened to bury them in the great crevassed mass below, then suddenly hurled them upward as through a wind tunnel, on wild rides that left them oxygen-starved, fighting deadly vertigo, trying to hold on in the cockpits of upended aircraft, while the gyros spun crazily on the dash. They flew on instruments because they couldn't see through the thick wool of cloud and darkness around them, and because their windows were blocked with heavy rubbles of ice.

A trip over the "Rockpile" might also include intense sunlight, though seldom from start to finish. Pilots prayed to avoid downdrafts and wind shears that could sweep them into the graveyards of other crews who got themselves blown off course. Out of radio contact, on primitive and faulty radar, shorted on fuel, they dodged the world's highest peaks, often barely clearing summits. No "learning curve" existed in the cruel surround in which they worked.

"If you were on the roster, you got on the airplane and flew it," Ned later remembered. "Clear weather was the exception. You went out there, put yourself in that metal cocoon and cranked up. Oftentimes I never saw the ground again till I landed in China. But I knew that under me were a lot of high, hard rocks."

The Hump was mean on airplanes, too. Men worked all configurations of hours in a day's twenty-four; planes could be in the sky again in an hour or less. "If they checked out, we flew them," says Ned, who only once refused to fly, when his magnetos failed to register on takeoff. He taxied back in a cloud of dark and belching smoke, and took off in another plane. "No big deal but you write it up." Like every pilot, he knew even a minor "deal" could bring down his plane.

The atmosphere of the Himalayas was an unforgiving habitat of stark killer weather, swirling typhoons, and eternal bizarre convolutions of wind that gave no quarter.

And that was before you took the enemy into account.

All of it grew into the story of an unparalleled war-within-a-war that got its own name.

The Hump War.

* * *

For a lively boy growing up in a small town in the North Carolina Piedmont, the dream to become a pilot could have begun with his dirigible ride in the early 1930s, on a visit to one of his grown-up brothers in Washington, DC. Leaving his mature sibling safely planted below, Ned eagerly peered out the windows of the great blimp, thrilled to look down on the earth as the clumsy ship maneuvered across the sky—his first experience of flying!

Purveyors of Roxboro's two drug stores smiled at the youngster with brown eyes and freckled nose, a shock of dark hair—son of a former schoolmarm and the aging owner of the town's hardware store—as he sat cross-legged at the magazine rack poring over issues of *Life*, *Time*, *The American*, *The Saturday Evening Post*, anything he could find containing a story about airplanes. Steeping his imagination, stoking his dreams, Ned knew the history of the Wright brothers' flight at Kitty Hawk in 1903, the nativity of modern flight there, the army's rapid interest whereby military aviation came into being. He read of the exploits of the Red Baron, Manfred von Richthofen, the Ace of Aces in the German Luftstreitkrafte (Germany's air corps during World War I), and of Charles Lindbergh's lonely flight across the Atlantic in 1927.

Before instrument flying, mail pilots would navigate their open-cockpit biplanes across the country guided by the Light Line, a series of towers with beacons that flashed a Morse Code signal to orient them at night. Ned himself experienced a throb of fascination at his later glimpse from the cockpit at this series of towers, by then in disuse, though still a striking sight as he gazed down from the darkened sky. The end of the 1920s brought the revolutionary break-through of the radio compass, the '30s, developments in twin-engine aircraft to fly travelers across the nation.

In his teens, Ned first heard of the retired Army captain and former stunt pilot Claire Chennault, who answered Chiang Kai-shek's call to China in 1937, and organized the flyers variously known as the American Volunteer Group, the Fourteenth Volunteer Bombardment Squadron, most famously the Flying Tigers. With other units aiding General Chiang, including OSS and commandos like Merrill's Marauders, the Tigers became flying idols, and they were led by a brilliant tactician. Convinced that Japan could be trounced from the air, Chennault advertised for volunteers. Whether or not a particular Roxboro boy dreamt of joining up, those who did—servicemen, soldiers of fortune, adventurers of every stripe—signed on ready to fight, ready to draw the generous bonus for every enemy plane they shot down, though not always so ready to venture near the tortuous airspace of the Himalayas, which caused numbers of them to turn around and leave as fast as they'd come. Those who stayed captured the hearts of the world, much like the pilots who flew the Battle of Britain in 1940, and inspired Winston Churchill to say, "Never before in human history was so much owed by so many to so few."

An age was dawning unlike any before. The heros were the men with wings.

By that recent spring of 1942, a new champion took to the skies, when on April 18, Lieutenant Colonel Jimmy Doolittle led sixteen B-25B Mitchell twin-engine medium bombers on the first air strike to penetrate Japan. Taking off from the USS Hornet in the Western Pacific, all the planes were lost, but most of the pilots and crews escaped, and Doolittle himself was flown out of China across the Hump, a feat that brought vital evidence to bear on the need for the route. In fact, some of the first materiel flown over the Hump was the high-octane fuel intended for Doolittle's planes. Even the downed raid was reckoned a success in the end.

Doolittle and his pilots hit five Japanese cities including the capital at Tokyo. His feat brought a surge to flagging U.S. morale, and proved, too, that Japan was not impervious to a strike by air, as many believed.

The Japanese unleashed a series of brutal reprisals. They slaughtered a quarter-million villagers and peasants in China, then on June 4, attacked American carriers in the Pacific, in the Battle of Midway. When the prized imperial First Fleet went to the bottom of the sea in the engagement, along with some of Japan's best aircraft, the writing was on the wall. Doolittle's raid was launching air power as a premier strike weapon of modern warfare.

Pilots formed a new elite that men wanted to be part of. There was a glitter, a prestige. "I had an aspiration to fly, and a basic interest in aviation. And as they joked," Ned's eye twinkles, "the man with the wings gets the girl!" Owing in part to their height, Ned and Deming aimed for the larger planes, perhaps like Doolittle's, despite a rate of fatality and loss on these aircraft that was extremely high compared to the smaller, maneuverable tight-fit fighters. The two decided to try for troop transport or perhaps the big B-17 Fortress or the B-24 Liberator.

Uncle Sam was not remiss to supply their opportunities. To a flood of conscripts, rail passes were issued, and in Miami they had free hotel rooms. "You didn't pay to go be killed," Ned quips. He never forget a conversation he struck up on the train with a man who limped and had a club foot. "I felt bad for him. He'd been refused twice for service, and was on his way to try again to sign up. That shows how strong the patriotism was running."

"Basic" in Florida—they didn't call it boot camp in the army—was no vacation. Recruits got disciplined to the military lifestyle, accustomed to guard duty, drills, lectures, rules. They saw films and studied history. A bit of excitement was stirred by some submarine sightings off the coast, from which enemy agents were anticipated to come ashore in rubber dinghies with intentions to blow up something. This, however, did not happen in

Miami. Fellows got issued uniforms but no weapons. They marched in formation across ball fields and parks, up the streets and down, as elderly vacationers sat rocking on the verandas and waved patriotically when they passed. Forty-five days later they got their next orders, Ned to Duquesne University in Pittsburgh, for a College Training Detachment, or CTD, a brief service academy arrangement at a college.

The CTD program—begun in January of 1943, with Ned one of the earliest enrollees—continued in the U.S. till July, 1944. CTDs functioned in some respects to regulate the movement of bright officer candidates along the pipeline. Afterward it was found that cadets who came through (some 150 CTDs were instituted at universities round the country) did better at pilot training and washed out less than those who weren't exposed to the experience. Ned marched with his fellow cadets across the university grounds, into class, in and out the dining hall, while across the nation conscription numbers swelled, and it became the task of the War Department to decide what happened next to each. The academic subjects intrigued him—history and military history, mechanical engineering, higher math, English and physics, courses to prepare him for the future in vital ways, where before he'd sometimes felt boxed into introductory college work with professors who struck him as monotone or out-of-touch. Flying was not taught in CTD at Duquesne, but cadets were taken up in Piper Cubs to be introduced to the experience and sensation of flight. Ned became the head cadet of the school, finished high, and wore the five stars of their commander.

"Though I was still a private," he clarifies.

Late summer brought news of a hometown friend a few years older with whom he'd grown up. "Bitsy" Bullock, whose family owned the town lumberyard, was shot down on August 1, 1943, flying a B-24 in Bucharest at Polesti, a major supplier of Germany's refined oil. Fifty-three planes and many irreplaceable lives were lost, and there was shock and sorrow in Roxboro.

In September, Ned left Duquesne for the Classification Center in Nashville, Tennessee, where further examinations sorted cadets into their future course. Those with highest marks would enter pilot training. Next came navigators, or a step below bombardiers, on successful completion of which they'd commission as lieutenants. If they washed out they got sent to the infantry or elsewhere, certainly not home. After testing at the top, Ned reported to Maxwell Field in Montgomery, Alabama, for the ten weeks of pre-flight that officially conferred the rank of Cadet Aviation Officer. With his head clipped GI style, he pushed through a rigorous stint of PT and survival, Morse Code, mechanics, military law, preliminary navigation and charts, aircraft recognition, firearms, and other subjects. He was issued a saber, and at the pay counter, saluted and recited his serial number to collect his minuscule private's salary. Ninety days later he got orders to Douglas Field, Georgia. He'd just turned twenty. Time to get down to business.

Time to fly the planes.

* * *

The airlift that came to be called a modern military miracle was undergoing a discordant and unpromising start. Plagued by bungling leadership, inadequate resources, an aboriginal operation in the worst sense, time and again the Hump looked doomed to fail. There were Japan's ferocious challenge to control the territory, China's blockaded state, Chiang Kai-shek's isolation in mountainous Chungking. And always, the horrific flying conditions.

In the unbroken savage Sino-Japanese conflict, what made compelling case for supporting the Nationalists at so great a sacrifice of life and resources for the U.S., was the massive contingent of enemy engaged on China's soil, soldiers whom Japan otherwise would have moved into the Pacific to fight and kill Americans. The rationale kept hopes for the airlift from faltering entirely. Yet at every step the project fell prey to logistical night-mares. Leaders at cross purposes created persistent moral and institutional blockades. Men and airplanes weren't the only ones to crash over the Hump. Egos clashed at every level.

With President Roosevelt standing firm on the necessity for the route, early in 1943, General Henry Arnold, Commander of the U.S. Army Air Force, had the War Department commandeer a couple of dozen DC-3 Dakotas, a civilian plane adapted as the C-47 by the military. It proved a tepid start for so hugely-con-ceived an operation. Few requisitioned aircraft even made it as far as India; as for planes on the scene, two-thirds stood ground-ed for lack of engines and parts, owing to the near impossibility of getting supplies where they were needed. General Arnold also sent General Louis Brereton of the Tenth Air Force, along with his chief of staff, General Earl Naiden, to whom fell the bedevil-ing task of trying to survey a flight path for the U.S. military across the forbidding range. Already, Arnold was demonstrating the aversion of American commanders to giving free rein in the Hump either to Claire Chennault, whom they regarded as a loose cannon, and over whom they had no control, or to the strapped Chinese National Aviation Corporation, called CNAC. Naiden took in the appalling scene and felt ill, perhaps no surprise given conditions. Concluding that twenty-five planes at most could fly the route, if even so many in the monsoons, he left China behind with the same impasse—little materiel crossing the range, the country's isolation intact, and for pilots, steadily climbing fatali-ties on CBI air runs.

It took a series of high-level conferences that year, first Casablanca, then Trident in May, Quadrant in August, Sextant in December, to focus serious international attention on the Airlift. Russia's Joseph Stalin and England's Winston Churchill joined Roosevelt and Chiang Kai-shek to parley. Where the China-Burma-India situation was concerned, each head of state seemed to pull like a dog at a bone in his own direction.

Chief of the Nationalist Party, or Kuomintang, since the death of its founder Sun Yat-sen in 1925, Chiang Kai-shek first had tried to work with the Communists and their leader Mao Tse-Tung, but by 1939, their views had split irreconcilably and they were locked in a cutthroat struggle for power. The Generalissimo's foremost ally at the summits may have been his charming American-educated wife, May-ling Soong. To marry her he'd consented to become a Methodist, and though his conversion did not make him a faithful husband, so dependent did he grow on her ideas, that her pretty face often appeared alongside his on occasions of statecraft.

May-ling's brother Soong Tze-Ven, Chiang's foreign minister, meanwhile traveled to the U.S., and met with FDR to pedal the feasibility of an airlift. An easy route, Soong smiled reassuringly, bestrewn with "comparatively level stretches." Not for nothing would it be said this smooth talker "won the battle for China in the corridors of Washington."

From the time of Japan's invasion, China's 20th Century social historian Lin Yutang, was a leading Asian influence on American scholarly views regarding his country. Skeptical of sudden change for his people, he advocated passionately for Chaing, a figure of courage and integrity, in his estimation, in contrast to the corruption of Europe's leaders. Lin spares the West nothing in its indifference to the catastrophic upheavals of the Far East. "The arrogant, unintelligent and unscientific Aryanism of Nazi Germany, the abandonment of Czechoslovakia by France, the stockbrokerly peace of Chamberlain in England, the triumphant cynicism of Mussolini in Spain and England's complicity in it, and the surrender of the League of Nations to Italy, give us a devastating picture of moral bankruptcy. The Chinese realize that if Europe is not willing to fight for justice in Spain, much less can justice serve as the principle of dealings elsewhere."

How does this pundit size up their Generalissimo? "The supreme chess-player of the Far East, and one of the greatest political chess-players of all time." A good chess-player is a cool player, he adds, "and this great enigmatic personality, whom I had watched rise to fame and power for ten years, could be inhumanly cool." He praises Chiang Kai-shek's stubbornness, a man "who knows what all this is about, and who views it as a twenty-round match, in owning that it is the final knock-out that counts." A leader on his guard, not quick to show his hand, Lin describes him, but a very human man "with human emotions common to every mortal; in particular he is a born last-ditch fighter.

"He fought a dozen major and minor civil wars, at times against formidable combinations, and had always come out victorious by sticking to his guns. He outwitted political adversaries and outmanoeuvred military opponents, so that they seemed always to be in the wrong. He bought and sold generals, had to work with a medley of politicians as he found them, with the bitterest cross-currents of internal politics . . . (He) had to tolerate evils until the time came when he could do away with them." [3] Lin's prediction? A strong dictatorial regime under Chiang Kai-shek, "different from totalitarianism of the Hitler-Mussolini or the Stalin type. The personal prestige of Chiang will become so great through his courageous and wise leadership that he will command a huge following and the loyalty of the nation. He has not renounced, and will not renounce, the democratic principles of (his predecessor) Sun Yatsen." [4]

Chiang's autocratic bearing belied to a point his position at the summit table. He was a supplicant. His men were exhausted, his resources sapped. Urgently he needed money, owing in no small part to a cadre of generals on the take whom he ruthlessly tried to root out and often had shot. Western leaders speculated whether he'd make it through the war at all. As their doubts grew, the Far East tumbled further in Allied priorities.

Still, it remained a complicated picture. Britain's patrician antipathy to having her imperial interests challenged by independence in China, or to losing her "jewel in the crown" India, sowed an Anglo-American split that gave rise to hair-splitting diplomacy for Roosevelt. Keeping Churchill appeased was a sideshow in itself. The Prime Minister's double-faced aim to end all tyranny in Europe, and at the same time preserve lordly colonialism and exploitation in Asia, could certainly have been called inconsistent. Chiang fortunately had been used to dealing with incongruity and duplicity for a long time. Skilled and subtle, he saw his wedge and used these high-level scuffles to his advantage. Aware that Roosevelt counted on him to step in and fill the Asia power vacuum after Japan's defeat, proudly he demanded to be "treated as a great power," if a future one. It could be bluntly said he had little else to bring to the table. Militarily his position grew more impossible by the day. He, too, confronted a dual war—on one side, the trained and equipped Japanese armies inflicting losses by hundreds of thousands, on the other, the rising Communists opposing him within China. Adroitly writing off the British for the time being, he played on America's guilt over colonial injustices, portrayed his country as America's "special burden," and worked to extricate greater promises of support from Roosevelt. When all else failed, he could exploit U.S. fears that he might forge a separate peace with Japan, despite his having already rejected several such offers from Tokyo.

Into this rambunctious arena two other combatants figured in a large way, though they did not parley at the summits. The first, General Joseph Warren Stilwell, was co-commander with Chiang Kai-shek in Burma till 1942, commander next of U.S. forces in CBI, supervisor of Lend-Lease there, and in 1943, Deputy Supreme Allied Commander under the British Vice Admiral Lord Louis Mountbatten. At present Stilwell burned with ambition to liberate Burma and reopen the Burma Road. It cannot be denied that this West Pointer and infantryman had been handed a thankless job, if not several. His responsibilities wove in and out a huge, tangled, unmanageable tapestry heretofore considered the purview of the British and Chinese, a territory that stretched from Karachi to Bangkok. Imbued to the marrow in the military thinking of the past, convinced that wars were fought and won on the ground no matter how measureless that ground might be, Stilwell adamantly rejected any expansion of air power as a waste of precious resources which, as he never ceased to argue, ought to go to him.

Called "Vinegar Joe," not necessarily with affection, Stilwell reeked with talent for underhanded dealings and epithets. He slurred his President as "Old Softie" or "Big Boy;" Chiang as "Peanut" and "crazy little bastard," Claire Chennault's air tactics as "medieval jousting." His letters and journals reveal a bitter, jealous man. His resentment knew no bounds whenever military supplies from the Hump reached Chennault. The Tiger's commander, however, was not the sole occupant in his galley of arch-rivals. He undercut Chiang Kai-shek at every turn, until his scornful, demeaning treatment and obsessive rage, grew so out-of-bounds that he considered having the Nationalist leader assassinated. To what end? No less than that this great and ancient country should devolve instead upon his own wiser Yankee head. [5] He'd made up his mind what was best for China, for the Orient, for America—and for himself.

By no surprise Stilwell failed to endear himself to Claire Chennault, the other big player in CBI. Chennault's influence over the Generalissimo remained consistently strong. Ever a sore point was that he didn't take Stilwell's orders, but ran his outfit and planned his tactics as he saw fit. General Chennault was aware of the contempt in which Stilwell held him, if not fairly contemptuous in his turn, but he depended on aircraft, weaponry, and fuel from the Hump, resources which often got diverted to other theaters by commanders along the pipeline, a routine snafu arising in part from the ending of the siege of Tobruk, Libya, in December, 1941, and the opening of the Middle East in '42, to the Allies, where much of the scarce materiel kept landing.

Chiang's wife May-ling had first proposed the air fighter

group. When she invited Claire Chennault to China, he quickly became her husband's most trusted military advisor. Commander of the Chinese National Aviation Corporation (CNAC) and of his own fighters, Chennault regarded the Burma Road, as well as the newer Ledo or "Stilwell Road," an insane squandering of manpower and resources needed to build airfields for the Hump. Instead he and the Generalissimo communicated a bold plan to Roosevelt. If the Airlift could be expanded, and if U.S. lawmakers would requisition additional planes, he and his pilots could knock out Japan's air power, mow down every obstacle, and clear the way for the Allies to sweep across the Pacific.

Roosevelt was intrigued, and then convinced. Chennault, who already was a general officer in China, received the two stars of a major general from the U.S. in 1943. Despite his recognized stature, or maybe because of it, he continued to be pushed aside in Washington by commanders put off by his blunt tendency to call a spade a spade. Vinegar Joe threw tantrums over the promotion, and forged ahead to deploy General Merrill and the Marauders behind Japanese lines. In December, 1943, he began his invasion of Burma, a campaign that would subject his men to more than a year of torturous, morale destroying ground fighting, as they struggled inch by inch across impossible jungle terrain. A study by Frank Sinclair the previous year (1942) already cast a cloud of doubt regarding Stilwell's "defeatist" attitudes, his stubborn prejudices, outdated tactics, and incessant clashes with Chennault, matters to which commanders might have paid some mind given the added friction Stilwell contributed to Chinese-American relations. Progress in CBI grew more hobbled by the day.

Never a friend to air power, Stilwell eventually (at the close of 1944) would achieve one critical gain for the Airlift, when his men liberated the airbase in north Burma, at Myitkyina. The same year would bring an unexpected turning of the tables in his own career when he lost his command and was removed by FDR from Asia. Another casualty of the Hump, at the bottom of the business lay his treachery to Chiang Kai-shek, and his shabby treatment of Claire Chennault.

* * *

The trains rolled into Douglas, Georgia, in early January, 1944, to disgorge two hundred aviation cadets—six new squadrons for Class 44G. With an upper division already advancing there, some 400 cadets taxed the facility to the max, a pace that would roar along till pilot needs leveled off. Primed to take to the skies in the service of freedom, the men hailed from every point on the U.S. compass and several allied countries. Named for Stephen A. Douglas, Lincoln's challenger for the presidency in 1860, the mid-size south-Georgia town with its farming economy of peanuts, corn, tobacco, and cotton, experienced a potent shot of adrenalin with the advent of this fully mobilized flight school, its 120 teachers, clerical help, building staff, mechanical and ground crews, and military commanders.

For Ned, the most memorable moment came with his first sight of the planes.

"They had these PT-17s all lined up, wing to wing. A long line of them, painted blue, yellow, bright colors. I think we had a bit of trepidation. But also some elation because we're going to move along. You come into Douglas, maybe you've never been that close to an airplane before, and all these aircraft are out there on the flight line. It's quite a sight.

"There's the military aspect. It's impressive. It affects you. We realize that's where our future's going to be, and it's all out there, all in a line."

A utilitarian scene of operational buildings constructed to minimize costs in a war-strapped economy, Douglas Field boasted a central tower for observation, though not for radio transmission, as radios were not used in primary flight. Two main air strips and four auxiliary fields, all sod, allowed for landing and takeoff in any direction. The time of year was as cool as Georgia gets, with weather often fine and clear, and the winter sky icy-cold in the open cockpits of the P-17. Cadets were issued leather flying jackets and trousers, fleece-lined for warmth, as were boots and snug leather helmets for their heads. Classes were stratified so that when the "big shots" moved on, those below took senior place, and another contingent arrived.

Even with its military austerity the base boasted a few amenities—a dining club for special meals with families or guests, a student newspaper whimsically named The Zombie, a bus to town for a rare evening out or movie at the Martin Theater. Pay was $75 a month, an additional dollar a day being deducted before payday for housing and meal allowance. Local cooks worked the mess hall, food was considered good, quarters reasonably comfortable. One of 56 primary flight schools operating around the U.S., staffed by civilian teachers and run by the military, Douglas alone survives, functioning now as the World War II Flight Training Museum, the most intact of such sites remaining on the nation's soil. A 1941 barracks stands to this day.

Prior to Pearl Harbor, primary flight training extended for ten weeks, but this was shortened to nine with military methods and discipline, and the urgent need for pilots. Not till late the following summer would the program's length be expanded back to ten. Ned himself remembers an intense, fast-paced mid-war program.

To begin flying was no guarantee to finish, and the numerous washouts were devastating for those who'd tested well, to discover they had little aptitude for flying, if not a fear of it. Even air sickness took a toll. General Cleo Bishop, a pilot himself, observes, "There was an extremely high washout rate, I think not so much that the people who washed out couldn't learn to fly, but they couldn't learn to fly fast enough." [6] Cadets were expected to make a first solo by 12 hours in the air; the most successful might do so in 8 or 9. In the mobilization of a vast country, the military had more men than time, and the quality bar was being raised, a rationale defended by Douglas's commanding officer, Major Thomas W. Bonner: "Do we want our standards lowered and elimination rates down, or elimination rates high and our standards high?" [7]

How high? By statistics nationwide, 28 percent of air cadets washed out at Primary level alone. Inclusive of levels through Advanced, Ned himself witnessed astronomical losses, later guessing maybe 60% of the fellows he encountered dropped by the wayside before the finish. Actually his reckoning is conservative. Writes General William Tunner, the last commander of the Hump, "the washout rate was high; seven out of ten officers failed to make the grade." [8]

Little slack was cut for mistakes. A cadet's ability to rebound from the unexpected formed one of the most vital aspects of Primary flight. For some, the specter of getting mustered out hung like the pall of Caesar's ghost. Air Force pilot Charles Watry titled his reminiscences about cadet training out West, *Washout*. Never to be undersold, a pilot's own temperament played a big part in this training, his ability to keep his head cool, draw on his knowledge in emergencies, and compartmentalize pressures and apprehensions till they didn't rule.

Aviation cadets' schedules were structured for a half day of classroom or ground training, with flying during the other half, a morning or afternoon of each. An hour of calisthenics, games, obstacle courses, and so on came at late day, and twice a week, a cross-country run. Core academics included an introduction to the mechanical systems of aircraft, cockpit and controls, meteorology—ever a concern—and other aviation topics, with writing and tests besides.

Legion were the American aviation cadets who took to the air the Stearman PT-17, "a tough little plane," Ned describes, "but it didn't have a lot of power." Developed in the '30s, the open cockpit, single-engine biplane had a canvas-covered fuselage and wings, the latter under-framed in wood, a body of welded steel, a propeller at the nose, and a geometric exoskeleton of supports. With a maximum speed of 124 mph, a wingspan of 32', length 24', some ten thousand were built by Boeing as war trainers. The plane's drawback was a tendency to ground roll, a mishap that landing pilots were ever vigilant to avoid, not always successfully. Durable and foundational with a seat fore and aft and controls at each, the PT-17 now is considered a classic, and a few still turn up for show flying. In their PT-17s, military pilots acquired the earliest fundamentals to survival in the sky.

"They threw us in pretty fast," Ned remembers. Issued his bunk, linens, and flying gear, he was soon in the air with an instructor. The initial ride served as an indoctrination to give cadets a feel for the controls, stick, and rudders, and to see how things looked inside an airplane and from over the wing of an open cockpit.

From here the pace never slackened. Their civilian teachers, called service pilots, faced perils of their own with inexperienced students and many planes in the air at once. Several had been born in the early century and learned on Great War era aircraft. Ned's own first instructor was an older man, who may have singled him out as a quick study, or maybe found himself struck by his personable and open manner. In any event, he offered what this particular cadet always met positively, and that was room to excel.

Sitting in the rear seat, Ned learned first to take off and land, then to fly the aircraft and start coping with stalls and other emergencies. Controls were available at both seats, with student and instructor staying in contact via a rudimentary talking tube called a Gosport, which shows in old photographs somewhat like a bent straw on their leather helmets in the vicinity of the ear flaps. The tubes being notoriously unsatisfactory over engine noise and technical glitches, pilot and student often communicated with hand signals and shouts.

The propellor of the P-17 had to be manually cranked to start the engine, and every morning began with the same process on the flight line. A couple of ground crews consisting of two or three men, would move up and down the line with a large leather sling attached to a rope long enough to bring leverage to the prop without danger of injury by it. The propellor's two blades first would be manually moved to a vertical position, and the leather sling attached. Then plane by plane, the ground crew called "contact," the pilot turned the magneto switch to "on" position, and they jerked the rope to swing the propellor and start the engine. It also was possible to use a crank, but the process still took two people and was harder than the leverage of a rope, so crews largely avoided it. The rope and sling didn't work every time either, but usually it did, and with their planes starting to purr, the pilots taxied out.

Crucial lessons introduced stalls. "How does it feel when a plane stalls? You have to feel it, and react to it." Ned remembers how without warning, his instructor would turn down the throttle and put the plane into a near-stall midair, leaving him to bring the engine back to life. "You're into the stalls very fast. You've got to be able to recover. Nothing's worse than going into a stall, losing flying speed, and not being able to pull out.

"A landing is sort of a forced stall, but in the air stalls can be caused by a lot of things, even ice on the wings. If you don't have enough altitude, a stall can be pretty fatal. So early on, you go up with your instructor and practice as he deliberately retards the throttle. Then you fall off, that is you tip down at one wing or the other, and start to spin. To recover you've got to stop that spin.

"You know you can't fly without speed. You're falling to one side and you'll go down spinning, so you have to apply rudder from the opposite side of the spin, then get the stick forward to attain flying speed. You pop the stick, which controls the two elevators and the ailerons. Add throttle and you have the flying speed to come up. Where men had trouble was near the ground, where you don't have the altitude to recover.

"If you completely lose the engine, you put the nose down and glide. Then you either recover the engine or pick out a field to make an emergency landing. The instructor may have you picking out the spot before he restores the power. You've got to know how to do it."

In a spin to the right, nose down, the student has to push the left rudder and the stick forward. Spin to the left, and push the right rudder, always opposite the direction of the spin. In another recovery maneuver, Ned's instructor would have him sighting out an emergency landing place, then on approach, he'd throttle the plane to jimmy what first looked to be a smooth landing, so they had to pull up again.

"You'd be leveling off, and he'd push the stick forward to make the plane bounce. Then you had to recover and land the aircraft. Hopefully. Coming from a bounce you don't have the flying speed you need, so you goose the throttle, burp it a little. You're hitting the throttle to level out the plane, and that winds up the propellor a bit." These ofttimes repeated lessons over the Georgia fields provided Ned's first measure of insurance against the rough and torturous mountain flying he was to do.

"You spin if you stall," he describes. "In an induced spin, the propellor is windmilling (i.e., idling) and as you lose speed, you start to fall."

He was barely into training, landing after another recovery from a hard bounce, when his instructor climbed from the plane, shrugged off his parachute harness, and turned to Ned.

"Think you can handle it now?"

Ned's reply was to climb back in, take off alone, circle the field, and bring it back.

"The best landing I'd done," a hint of remembered pride in his voice. "A grease job."

His instructor had him go up and perform three landings in succession, the number he had to satisfy to fly solo. The first in his class, Ned soloed after less than seven hours in the air.

Marching back to the flight line, he would've been the focus of many eyes as he moved his goggles forward on his helmet, the protocol being that goggles must be worn backward on the ground till a cadet qualified to solo. He got his name recorded by dispatcher and townswoman Nola Fitzgerald, who peered through the wicket, beamed, and broke the silence.

"Well, good for you!"

A pamphlet for Douglas cadets underlines this and assorted other advice, including that men not neglect to get a haircut once a week. "Cadets will neither wear wings on their garrison caps nor wear flying goggles on their foreheads until they have soloed from this field." [9]

Ned's teacher cut him loose and from that time he worked with an instructor about six months younger than himself, J.M. Woods. "A kid. We got along well. I don't know where he learned to fly, must've taken lessons while I was in college. Could be that Uncle Sam caught up with him and made him a flight officer down the line." It says something of a cadet, that he called a teacher near his age a kid, yet apparently didn't see himself as one. Another instructor at Douglas, Ben Epps, though not so young (closer to thirty) went on to be a pilot in the Hump.

Check rides given by civilian or sometimes military pilots, occurred at intervals. There were cadets who froze during the ordeals, knowing failure to get everything right could wash them out. Ned and doubtless others viewed them as fairly routine, and moved along to what came next, which for those who survived to late February, meant acrobatics to hone their skills.

"Stalls and spins came early so the student would know what a plane feels like if it's about to stall. Loops and acrobatics were maybe two-thirds of the way into the program." In the recollection, Ned wouldn't be speaking of stunt flying but of handling a plane competently. He learned the slow and fast rolls, the more intricate loops, the quirk at the corner of his mouth doubtless lifting as he executed these feats successfully across the bright Georgia sky.

"You do a roll with power to keep the plane flying, either a slow roll or a snap-roll, just keep the nose up steady, as you work the stick and rudder together. Slow is done piecemeal. If it's a snap, say to the right, you kick your right rudder and pull the stick back. A slow roll can be harder than a snap." There's the Immelman, a half-loop that turns the plane back on itself in the opposite direction. To execute the maneuver, the pilot goes up in a rapid climb pulling back on the stick, moves through a half-loop, then a half-roll, as he levels out and exits at maximum speed. Also called a roll-off-the-top, the name is associated by some with the dog fighting tactics of World War I, but rightly understood the maneuver, like its steep-climbing sibling the Chandelle, was part and parcel of aircraft mastery. Fighter pilots like Claire Chennault always knew the dog fighter's best friend was an expeditious dive.

Douglas ended before Ned was wholly satisfied with his Immelmans, but he had no cause for shame, he'd finished well. His military record notes 145 landings at primary level, a number that may not be complete. In March, came the farewell parade, banquet and graduation ceremony—instituted by the previous year's 43G class—to which the public was invited and Douglas's leading citizens turned out. Ned had the award for first solo but no time to celebrate. His orders were cut for Basic Flight at Cochran Field, in Macon, where he was to meet up with a plane known as the Vultee Vibrator, do his first flying on instruments, and polish his acrobatic work.

CHAPTER 2

"NOOKS"

JIMMY DOOLITTLE'S AIR STRIKE ON JAPAN MADE HIM THE nation's hero, but pilots the world over stood in his debt for the pioneering work he did in instrument flying. In 1929, Doolittle used inventions he developed for the cockpit with radio beacons to establish the position of an aircraft. By instruments aviators could control their planes when they couldn't see the ground—at night, in storms, over water, or in cloud cover, all the conditions that could disorient them and distort their senses. At the second level of flight training, called Basic, at Cochran Field in Macon, Georgia, Ned first worked with this innovator's formidable achievements.

Macon cadets practiced instrument-guided approaches and landings on the ground inside Link Trainers, a clumsy apparatus Ned describes. "Sort of a box on a platform, it had stubby wings, you went up the steps, climbed in a cockpit, and started the pneumatic. The thing had a throttle, a stick and rudders, and an instrument panel. You put on the earphones and a technician sent you the signals. It was early, very rudimentary. You'd never do that today."

Cadets had to master a higher level of Morse Code—sending, receiving, copying, sometimes through earphones, and be proficient at reading light signals. In navigation seminars they were assigned departures and destinations. They planned their flights using compass headings and maps, and overflew their points in their Vultee B-13s as spotters on the ground sighted them. Ned retains a graphic memory of his single worst flyover. After slowing to cross his point, he advanced the throttle, causing an inexplicable sooty billow to surge from the exhaust stack below the cockpit. Alarmed about his engine, and to avoid banking into the smutty froth, he turned the plane opposite from the direction he'd mapped, and the spotter wrote him up. Landing with a dark plume belching in his wake, he found a cluster of military instructors waiting grimly on the flight line. No one smiled at his approach.

"They weren't sympathetic. For punishment I had to walk laps wearing my parachute."

No demerits given, so it could have been worse. The Vultee BT-13 "Vibrator," having a self-starter required no cranking by ground crews, but the cockpit Ned remembers was a rattletrap. "They used to say when you revved up, the panel vibrated so much you had to jump up and down to keep in sync with your instruments." A low-wing cantilever metal aircraft with tandem seating fore and aft and a sliding canopy that joined at the middle when hooked together, it was heavier than the fabric PT-17 he'd flown at Douglas, and a lot of pep issued from the single 450-horsepower engine built for acrobatics and rapid handling. Ned maneuvered through loops and rolls, the slow rolls, which in some respects he found more challenging than the snap rolls with their faster operation of stick and rudder.

Training was rigorous, and men had to stay focused. They tackled academic subjects including weather reconnaissance, clouds and fronts, mechanics, aircraft maintenance, engines and engineering, all of which underline Ned's point, "You had to know about the plane you were flying." By the time they reported to the mess hall for dinner, they were dog-tired, though in this regard the U.S. Army Air Force didn't let them down. "I'll have to say all in all, the military did pretty good with food." It didn't hurt that they were young and usually ravenous.

Basic brought pilots their first experience in night flying. "You worked out your flight with compass headings and maps. They'd tell you where to go but you had to plan the navigation, and you had to overfly your destination so the spotter on the ground saw you."

How important are instruments? Ned is blunt. "You can teach a monkey to fly an airplane, but to fly it safely you have to know how to fly by instruments." Compass and altimeter, the RPMs, manifold pressure, artificial horizon, these readings are things a pilot monitors constantly, because regardless of what he sees or hears, difficult flying can demobilize his human perceptions. He's got to learn to trust his dials. What an inconceivable testimony to the bureaucratic blind eye, that pilots were being sent into the Hump well into 1944, without having been taught instrument flying.

Stalls and emergencies they reviewed again and again. "If you can maintain flying speed, then you can control the aircraft. If all your engines quit, you can glide. But if the plane goes below flying speed it's going to stall, and if you have no altitude, you can't come up to speed. You have to get speed to come out of a stall. If you have altitude you can maintain your speed by descent—by putting your nose down and setting up a glide. But if you have no altitude, you're in a worse-case situation. You've got to put it on the ground while you have some flying speed. You've got to hit that sweet spot. Too much speed and you've not going to get it on the ground. You'll overshoot the runway."

How many times would Ned repeat these lessons with the cadets he'd soon be teaching to fly.

It was coming together. Military instructors—teachers being military at this level and above—critiqued and graded cadets on piloting and academic work. Some failed to master advanced Morse Code and washed out. Ned, however, hit his marks, and on May 29, crossed Georgia again to report for Advanced Flight Training at Moody Army Airfield, in Valdosta.

* * *

Between April and October, 1944, during Ned's stint at Moody Field, a worldwide war raged without letup. June 6 brought D-Day to the Cherbourg Peninsula in Normandy, and the invasion of Europe under General Eisenhower.

In the Far East, Japan launched a brutal land offensive, code-named Ichigo. Intent on thwarting further bombing of the homeland by American B-29s, and on purging the bottommost guts from Chinese Nationalists, Japanese imperial troops mowed down resistance from Manchuria in the north, to French Indochina in the south. They destroyed Chennault's air bases, and had they ended as they began, could have pushed up the Yangtze valley into Chungking, done away with Chiang Kai-shek, and ground his forces into the dust. Anyone harboring the notion that China constituted a backwater war could avoid the reality no longer. Of two million Japanese troops occupying the country, two-thirds of a million were committed to Ichigo alone.

So where was Churchill? Still fixating on Singapore, "the only prize that will restore British prestige in this region," he intoned in the manner that was his alone.

Some blamed Claire Chennault for not halting the terrible aggression. Stilwell was one. The question is, how could he have? Though he worked with Vinegar Joe in superficial respects, Chennault mapped out his own operations, and it was understood his command eventually would expand to the Hump. Stilwell as yet retained critical powers, however, including the requisition of supplies flown in, and in this, as in all, he was not shy about throwing his weight around.

Which leads to another question. How important was the Airlift by early 1944? In *Thunder out of China*, authors Theodore White and Annalee Jacoby cut to the marrow. "The bottleneck for all the conflicting ideas, strategies, and ambitions in China was the Hump. For two and a half years the only contact China had with America and the Allied world was the fantastic airline that crossed the spurs of the Himilayas from upper Assam to the plateau of Yunnan. The Hump was the key to all politics in China. Stilwell, Chennault, and the Chinese government locked in bitter dispute over how the tonnage should be distributed." [10]

The Hump, a pilot's nightmare, had expanded to a power struggle at the highest levels.

Ordered to get vital materiel to Chennault, Stilwell didn't come through till October. By then Ichigo had mopped up every Chinese airfield, ruled the Manchuria-Hanoi corridor, and isolated Chiang Kai-shek in Chungking. It was a survival and evacuation scene that May. The entire Nationalist 6th Army was airlifted to Chihkiang to re-secure that single base. By now, two million of Chiang Kai-shek's soldiers had been killed, in numbers that continued to mount as Japanese troops laid waste all in their path. Civilian casualties were uncountable— tens and hundreds of millions, murdered, starved, tortured, fleeing as refugees.

Urgently Chennault petitioned for planes. His meeting with Roosevelt in Washington before Trident had been positive and raised his expectations, but any progress in China's air war was soon eviscerated by Stilwell's antagonism. By late October, 1943, the dearth of materiel had nearly reduced the Nationalists to the death throes. The struggling Hump Airlift needed to be mobilized rapidly into the rescue effort. How to meet the challenge? Night flying was introduced over the Himalayas.

Results were nothing short of horrific. In the first 60 days, 66 crashes occurred on the route. Crews went down and were killed to a man. Lack of training and experience by flyers contributed to the toll, but aircraft failure, enemy fire, and appalling weather all factored in. General Hoag writes ominously to General Arnold, "With the experience level here, we are going to pay dearly for the tonnage moved across the Hump."[11]   As if they weren't already.

Acknowledging that Chennault's fighters had to be supplied, Arnold brushed aside Hoag's dire forecast and ordered more planes to fly, and the runs increased, but as before, orders, schedules, and deliveries of supply shattered up and down the chain. The more things broke down, the more morale hit rock bottom across CBI.   Pilots who endured jungle ailments and mounting fatalities wondered why they should keep paying with their lives for supplies likely just to end up on the Chinese black market. By contrast, they never blinked an eye about giving to the last measure to help "Old Leatherface"—Claire Chennault—whose flyers' spirits were about equally in the tank as crashes and mishaps in the Hump tied their own campaign in knots. It was a symbiotic relationship between Chennault's Fourteenth and the Hump, for in those days, the task of defending the route also fell to Chennault.

There remained Stilwell's imperious, chronic deaf ear, a drag line no one seemed able to remedy. Of Chennault's multitudinous problems, many originated with Vinegar Joe.   Chiang's envoy to Washington, TV Soong, renewed his plea to Roosevelt to focus on getting resources over the Hump to Chennault's air war, finally threatening the Chiefs of Staff that Nationalist China would sue for a separate peace with Japan unless aid reached them.

Roosevelt made more promises. Chennault expected them to be honored. Stilwell obstinately argued for a Burma land campaign. Any increase of air power against Japan, he insisted, would only provoke renewed attack. Chennault shot back that China would collapse before that happened, whereas with a couple dozen more planes he could dismember the Japanese within China's borders, save Chiang Kai-shek's government, and do enough damage all around to allow Americans to advance across the Pacific in an unstoppable wave. In fact, air power was a weapon the U.S. could have used to huge advantage on the China mainland, but Stilwell only wanted to fight on the ground. He didn't like airplanes, he didn't like Claire Chennault, and he harbored a murderous death wish for the Generalissimo who'd been fighting for his country since the early century. Yet as Stilwell himself was to find out, one cannot treat a respected elder leader with treachery and disrespect without facing the consequences.

Stilwell got his jungle war in Burma. But even the British finally repudiated him, and that was the tipping point. It was said his angry showing at Trident, his chip-on-the-shoulder behavior at having his views questioned, ultimately lost his support. Chennault, by contrast, impressed FDR. He spoke effectively for Chiang Kai-shek and in any event, his reputation for straight talk preceded him. It was being recognized, too, that little was possible in China without the Airlift. Opinion at the top was veering around to an invasion of Japan to be launched from the western flank, that is from inside China, rather than by jumping U.S. troops island-to-island across the broad Pacific. Yet to do this, China had to be kept viable, and kept in the war.

Washington scrambled to order fighter planes for Chennault and more aircraft for CNAC, with materiel and supply in quantity to support them. Then as fast as new quotas were set, they broke down. "Everything seems to go wrong," Roosevelt writes with building frustration to General Marshall. "But the worse thing is that we are falling down on our promises every time."[12]

On the eve of Japan's Ichigo offensive Stilwell launched his drive into Burma. At the time, he operated with so sure a sense of his own indispensability and autonomy that he felt any objections from the humiliated Generalissimo likely would erase Chiang from the game and set up himself, Stilwell, as cock of the walk. Chiang Kai-shek could "go to hell," he said, the sooner the better. Meanwhile news from CBI worsened by the day. Writing for *Life* magazine, on September 11, 1944, war correspondent Theodore White flung raw reality into the face of U.S. leaders.

"The Hump," he writes, "is a line drawn across the eastern Himalayas and the forest of Burma by American blood and courage." Later that year, Stillwell would secure the last of Myitkyina (pronounced "Mish-a-naw) unwittingly aiding air power in CBI by capturing this major departure point of Japanese fighters launching against the Hump. At last big American planes would be enabled to fly the narrow neck of the southerly route they desperately needed, for no aircraft then could reliably or consistently handle the higher, more hazardous northern passage.

Stilwell by now had shot himself in the foot. Sick of his surly behavior, Chiang Kai-shek demanded his recall. Roosevelt replaced him with Major-General Albert C. Wedemeyer, who set up the 24-hour-a-day Hump timetable that would remain. Born now on the Top of the World, the age of airlift began with an operation destined to be the most massive of its kind in history.

An embittered and caustic man as he left CBI, Stilwell wrote in his diary that October, 1944, "Let me out of this odorous sewer, and I'll nevermore shovel manure."

* * *

Valdosta sprawls along a frying pan of topography more or less where Georgia starts running to catch the Florida state line. During Ned's stay there, 29 May to 4 August, 1944, temperatures topped 100 degrees, and even flying gear felt sticky. A hub of commerce in tobacco, turpentine, pine lumber, and pulp wood, the town of 16,000 burgeoned with the bustling military presence of the Moody Army Airfield, christened in1941 in honor of test pilot General George Putnam Moody. In January, 1943, the War Department activated there the 29th Flying Training Wing for advanced flight, to prepare pilots to fly transports and bombers.

Back at the time of his testing in Nashville, Ned and other qualifying cadets were given opportunity to specify their prefer-ence as to whether they wished to fly multi-or single-engine air-craft. Orders to Moody validated his choice, though this was not always the case and functioned, too, with demographics and need. At the advanced level aviation cadets usually flew the Beechcraft AT-10 Wichita for multi-engine (the AT-6 for single engine). Developed as a wartime trainer, the AT-10 was framed in wood and plywood to free up scarce war materials for operational aircraft. The twin-engine plane accommodated two men in the cockpit, measured 34' nose to tail with a wingspan of 44', cruised a bit shy of 200mph, and was surprisingly sleek in appearance considering how much of the body depended on woodworking and cabinetry skills.

"Easy to fly when you learned it," by Ned's assessment. "You didn't have to manhandle it, but being a tail-dragger it could be tricky to land, particularly in a tail wind." As a point of interest, a few months after his stint at Moody, the AT-10 was replaced by a larger twin-engine, the BT-25 Mitchell (T in a plane's designation signifies its use as a trainer) a changeover that added five extra weeks of operational work, and also put to use surplus Mitchells, which were used in every theater of World War II. An advantage of the '25 was the nose wheel which made it easier to land and taxi. A disadvantage was the lack of autopilot. When Ned himself eventually flew it he felt it "handled a little heavy."

For Class 44G (G signified the August completion date) several hundred advanced aviation cadets poured in. The 68 days in which they trained were fast-paced, and given the shortage of military pilots to instruct them, no one expected to be coddled. Cadets passed their final tests on instruments. The reverse side of a Form 5 from the War Department notes that Ned Thomas qualified on July 17, 1944, "in compliance with Army Air Force regulation," undersigned by his check pilot, Lieutenant WM Quinlan. He was authorized the white card that qualified him for instruments, to be replaced by a green card after a thousand hours of flying. Once a pilot had his "green card" he could sign his own flight plans, a protocol that doubtless got shortcut from time to time under trying war conditions.

To go from single to multi-engine aircraft, with more instrumentation to master and more tasks in the cockpit to look after, constituted a quantum shift in training. The AT-10 had a retractable landing gear, the first cadets had worked with. They did more night flying and more navigation, "really honing those skills." Sometimes Ned's instructor would pull back on the throttle and effectively shut down one of the engines.

"Then you'd have to trim the plane up for single engine flight until he restored the engine. We'd think, why mess around with fate when you've got two good engines? But it was one of the best things we learned to do." In future flying Ned lost an engine only once and counted himself lucky. "You can land a plane with just one engine, but it takes a lot of work."

The men flew with fellow cadets as well as with instructors. They mastered the precision ariel choreography of formation flying that can seem to defy all laws of nature when viewed from the ground. Ned enjoyed this aspect of the work.

"It grabs your attention when you try to put the wing of your plane in alignment with the next. But you get into an altitude and coordination, you're flying in the same air currents, the same speed, and formation flying really isn't that difficult. You just stay steady, allow the others to form up with you, it's not that big a deal. Of course, you have to keep your eye open. Otherwise you're going to chew up somebody."

What about time off? "I don't remember any time off, except Sundays." A day to worship, write letters home, make up sleep, it was the only break the men had.

The big moment came on August 4, 1944. The aviation cadets of Class 44G moved from cadet status to officer and received their commissions. A document dated 26 July and signed by Major General Butler from Headquarters of the Army Air Forces Eastern Flying Training Command, Maxwell Field, Alabama, contained the separate order rating the men as pilots. Promotions and ratings were commemorated in a ceremony in which Moody's cadet corps, some three hundred strong and hailing from every region of the country, marched in formation and passed review. In the auditorium, men received their commissions as second lieutenants, or the rank of flight officer. By custom, the first enlisted person to salute a new officer got a dollar, and there were soldiers waiting outside for the newly commissioned lieutenants, ready to snap to attention and with a wide smile pick up enough money for a night out or a movie.

Then the men got back into formation and headed for the parade grounds to get their wings, a moment Ned remembers as the "real occasion for the participants." His widowed mother drove from North Carolina with one of his hometown friends to share the event, and remained after the ceremony among the graduates, many of whom had no family present, and who flocked around for her to pin their new wings on their chests.

The so-named Special Order of the day sorted members of the class into their innumerable next assignments. Ten of the new lieutenants were selected to report to Randolph Field, Texas, for Central Instructors' School. These men had excelled sufficiently, and were judged suitable in temperament, to serve as instructors for the next pilots coming through. Ned Thomas was one of the ten. Meanwhile he had two weeks of vacation, his first break from training since he'd begun, and he was heading home to Roxboro.

Eighteen grueling months had rolled by without letup, since he'd reported to Miami, a college student and raw recruit. Today he wore the bars of an officer and the wings of a pilot.

* * *

Neighborly small-town Roxboro, North Carolina, was as good a place as they come to keep an energetic growing boy's mind and body working full tilt. The family's frame house on South Main Street, with its high-ceilinged rooms and a generous porch, had a large sunny kitchen where Ned winged after school to Rebecca, the cook, who kept a platter of warm fried pies waiting. The spacious back garden supplied kitchen produce, his favorite fig tree sagged with fruit in season, and shady pecans in front dropped their bounty every fall.

Having five boisterous sons to raise, Ned's father, George Washington Thomas, wouldn't have expected a worry-free life. Born in 1871, at an old farm called The Meadows, one of eleven siblings, GW as people called him, showed an industrious bent. As a young man he'd added "Washington" to his name for distinction in business. Ned, his youngest, saw a loving if distant parent; only his father's letters reveal the burdens he carried, as well as his painful struggle with spelling and grammar, for his formal schooling ended at third grade.

"My father was already old when I was born," Ned describes. "He was a Nineteenth Century man. He grew up with wagons and mules and horses. But he was a real entrepreneur."

This leading citizen built and owned the town hardware store and brick building it occupied across from the courthouse. He rented out offices upstairs, and leased to a druggist the corner portion where a soda fountain with chrome valves dispensed fresh fizzy drinks. The cavernous basement faced a tobacco auction and warehouse across an alley, where before Prohibition, a saloon flourished, and GW permitted prizefights on Saturday nights, tacitly smiling on the betting and back-slapping which might otherwise have been presumed nonexistent in an upright church-going kind of town. He also traveled to Kentucky and Tennessee to purchase mules at auction, some hardly broken, which he brought back on the freight train and sold after harvest time when farmers had cash on hand. Big, even-handed, and steady, GW enjoyed his youth as a time for sowing his oats, but the fighting and horse trading were viewed as manly behavior by the rough countrymen with whom he did business. On one excursion, to St. Louis around 1920, he rolled off an astonishing and unexpected cargo—a gleaming Premier Touring Car which had been custom built for America's baseball hero Babe Ruth (traded in December 1919, from the Red Sox to the Yankees). His eldest son William, seventeen years' Ned's senior, who later made off with the thing, praises it in superlative detail:

"It was maroon with a cloth top. When you opened the rumble seats it could carry eight passengers in comfort. It had oak-spoke wheels and cost $5600, the most expensive car in the world then!" These big convertible touring vehicles, which Premier made from the early century into the 1920s, featured Cutler-Hammer electric gear shifts, a Lycoming (later manufacturer of airplane valves) six-cylinder overhead valve engine, and in this instance, a specially built radiator said to cost $111 extra. Such a stir did the car create, that GW parked it out on the square between his storefront and the courthouse, where people flocked to have a look. It was being stored in the family barn by 1927, when brother William and few local fellows appropriated it with plans to drive to California. Loading the rumble with elongated bunches of bananas—one was a grocer's son—they waited for GW to ensconce himself at his private seat in the movie house. Ned lay innocently asleep in the large back bedroom in his little cot. "I looked down at him," tells William, "and said, 'So long, little fellow.' Because I knew I was getting ready to do something real bad." The comment, that elicited shouts of mirth in later years, would have made no one laugh at the time. The fellows drove the car here and there, even across a flowing riverbed someplace, till events intervened, often involving girls they remembered back home. William himself got as far as the coal country of Logan, West Virginia, linked up with some cousins, sold the car, and went working in the mines till his father's shock had time to abate.

GW proved himself unconventional in his marriages, each to women from South Carolina, the first of a family named Marcus. Though it was unusual for a small-town Methodist to court and marry a woman of Jewish lineage, he did so and they had three sons. After his wife's death in childbirth, someone else was bound to catch the eye of this mature man of business—this time their new schoolmarm from Gaffney, twenty-seven-year-old Anna Rossie Pridmore.

Independent Miss Rossie had a college diploma, unlike most women of her time, and was pretty, too, if a bit imposing with spectacles parked high on the bridge of her old-time hump nose. The couple married in the front parlor of the house as GW's three sons watched wide-eyed from the landing. That afternoon the bride returned to Gaffney to break the appalling news to her Baptist family that she'd married a Methodist ("forsaking her religion and her God," someone said) though they had to know she'd made herself a fine match on the whole. Her teaching days may have instilled the fortitude she needed to mother three half-grown boys, and she bore two sons herself, first Ned's brother Pridmore, then on September 16, 1923, Ned.

Called Nooks (rhymes with "Lukes") by everyone, Ned's nickname stuck. He gave a special name of his own to his mother, Muir he called her, which is pronounced "Muh" and rhymes more or less with "fire," which they call "fah," because that is how the language works in Roxboro. "I was a mama's boy," he grins. Years afterward, when he posted at the American embassy in Paris, he greeted a visiting fraternity brother who'd grown up in the house next door, Dr. Don Bradsher. Hearing Ned give someone directions in rapid French, Don turned to his wife and whispered passably loud, "I can't believe that's old Nooks over there speaking French."

Ned experienced the scrapes of boyhood and stayed busy. At the radio set in the next-best parlor, he followed the serials: the Lone Ranger, Buck Rogers, and Terry and the Pirates, with airplanes and a Dragon Lady villainess, set in China against an unnamed aggressor (Japan). His mother renewed his subscription to his beloved Boy's Life Magazine. He ran down Main Street to school, and came home at noon for dinner, which at

Miss Rossie's bountiful table meant fried chicken, brisket, cro-
quettes, or country ham, a fleet of side dishes, and pie, shortcake
or homemade ice cream for dessert. Evenings were vegetable
meals but all were served with fresh hot biscuits, thick gravy, and
from his mother's well-stocked pantry, homemade pickles of
cucumber, watermelon or peach, her relishes, chow-chows and
succotash.

"I could spit for a country mile," may have been the one boast
of his childhood, owning to a gap between his front teeth, which
Roxboro's dentist closed with a gold band and clever corner
screws they tightened at intervals. His sole health crisis, a punc-
ture by a rusty nail, was dealt with at the rear of the drug store,
where they poured iodine into the wound till Dr. Love could
arrive to inject the tetanus serum. His mother stayed up several
nights bathing him to bring down the fevers and combat the
hives, all preferable to lockjaw.

Saturday afternoon brought a cowboy show to the movie
house. Tom Mix, Hoot Gibson, Bob Steel, and Ken Maynard
were favorite stars, and there would be a serial, as well, "sort of a
thriller that would end with the lady on the railroad track, or the
cowboy and horse going off the cliff, till next week, and they'd
stop the action right before. People just ate it up." All the boys,
young, older, black, white, played football in a meadow behind
the Winsted house on Lamar Street, or gathered to enjoy a fight
when one of brother Pridmore's friends recruited an opponent to
challenge Ned. No one flinched, and the two swung around till
a winner was declared.

"It was all in good fun, it wasn't malicious. We just flailed at
each other."

In their Model A Ford, Ned accompanied his father to their
tenant farms to stock up on hams and produce. Farmers would
hoist him up on the mules, and with firm grip on a big collar, he'd
spur along the sturdy handful of beast beneath with his stubby
high-top shoes. They stopped at The Meadows, whose big house

pre-dating the Civil War by a decade or two, was imposing but needed work. There wasn't much to worry a boy whose family prospered reasonably, and whose hometown was of sufficient size to support a hotel, a bank, furniture and dry goods stores, and a remaining cotton gin, his father's having burnt to the ground when he was small. People rode to town on horseback and in wagons. They brought tobacco to the warehouse, cotton to the gin, and auctioned their crops. Agriculture governed the rhythms of life, though manufacture came with a fabric mill for automobile upholstery, built by the family of Ned's friend John O'Brian. John was 4F at the draft on account of his deafness: he'd saved the gas cards to drive Miss Rossie to Macon, for Ned's commissioning.

In the U.S., in 1933, Franklin Roosevelt's inauguration brought the New Deal, the creation of the Tennessee Valley Authority, and electricity's arrival in rural America. During these years Adolf Hitler ascended as chancellor of Germany, and Japan withdrew from the League of Nations. By his teens, Ned himself was following broadcasts by Edward R. Murrow from Europe, and the news in Roxboro's Courier Times: Hitler's elevation to Fuhrer. Germany's, then Italy's withdrawal from the League of Nations. The fall of Nanking, Canton, and Hankow to the Japanese. In 1937, the Nazis dismembered Czechoslovakia. Two years later, Germany invaded Poland, and World War II officially began. By now the airwaves were swollen with war.

"Everyone knew war was coming. France and England told Germany that if they invaded Poland, it meant war. They'd let Hitler have Czechoslovakia as appeasement. We started Lend Lease to Britain. Rationing came for food and gas."

In the lower reaches of his father's hardware store, the wide double doors would be swung out, and here Ned worked with the men to fit together the flue pipes for North Carolina's abiding tobacco barns. "I'm sure I wasn't overworked," he comments. "And I'm sure I was paid what I was worth. Which wasn't much."

The metal arrived in 3 x 12-foot sheets. At a table with a push-lever at one end to move the cutter, and using foot power, Ned helped slice the pieces to the required length, forced them through a hand-cranked roller to shape into cylinders, and hammered the ends before they went into the reducer that crimped one to fit into the next. Lengths got riveted together by hand to fit varying requirements. Heading home for dinner at noon, he might meet up with old Bessie Hester, the nanny who'd watched him protectively through his infantile years, and who of all women excepting his mother, he loved best. From his pocket he'd fish up a nickel and beg her to stop and buy herself an ice cream cone. By his return the farmers with their mules and wagons were pulling into the alleyway to load their pipes. Everyone talked of the war.

The Thomas Hardware Store carried everything imaginable to support an agrarian culture—kegs of nails, drums of linseed oil, coal scuttles, oil stoves, chinaware, buggy whips, firearms, ammunition, and about any tools a farmer couldn't readily make. Buckets dangled from hooks in the beadboard ceiling, alongside big padded collars for plow mules, leather halters, and leads. As in most places of business, there was an old spittoon. A newer innovation came with the hand-operated elevator GW installed. Any scrap left over from the pipe work would be loaded onto a wagon and taken to be weighed and shipped away to Japan, for in the days before the attack on Pearl Harbor, the U.S. sold the greater part of its scrap metal to the Japanese.

"Then it was all shot back at us," Ned quips ironically.

Menacing world events were superseded momentarily by a shock at home. In April, 1940, GW Thomas died, leaving Ned at sixteen, to mourn a parent he felt he'd hardly had time to know. In the weeks surrounding his father's death, Germany successively invaded Denmark, Norway, Belgium, the Netherlands, and Luxembourg. Neville Chamberlain resigned as prime minister of Britain. Winston Churchill succeeded him. By June, Hitler occupied Paris. France surrendered; and in July came the air war called the Battle of Britain.

Ned's widowed mother struggled to keep the family business afloat. It could have been an age since the summers of flue pipes, of smoke moving lazily through their shafts, of fireboxes stoked with live coals to spread a slow trek of warmth across the tobacco barns, and offset the Carolina humidity. Of big leaves dangling to cure from poles laid across the beams, while nearby, men gathered to pass the jug, feed the crackling flames, and baby the addictive foliage that was their surest promise of cash in lean times.

Finishing Roxboro High's academic track in three years, Ned left for Wake Forest College, in the fall of 1940. In his second year, on December 7, 1941, the Japanese attacked Pearl Harbor, Hawaii. Clustered round the radio set in the Pi Kappa Alpha fraternity house, the fellows listened to President Roosevelt's speech.

"Yesterday, December 7, 1941—a date which will live in infamy—the United States of America was suddenly and deliberately attacked by naval and air forces of the empire of Japan." No one stirred as their President named other Pacific outposts the Japanese had struck. Hong Kong, Malaya, Guam, the Philippine Islands, Wake and Midway Islands. "The facts of yesterday speak for themselves. . . . Hostilities exist. There is no blinking at the fact that our people, our territory and our interests are in grave danger. With confidence in our armed forces, with the unbounding determination of our people, we will gain the inevitable triumph, so help us God. I ask that the Congress declare that since the unprovoked and dastardly attack by Japan on Sunday, December 7, a state of war has existed between the United States and the Japanese empire."

Draft boards across the nation hit the ground running. Wake Forest organized its first ROTC program, and admitted a few women to help fill classrooms emptying of men. The U.S. sent military personnel to the college to study accounting and finance. Every able-bodied younger man was headed to war.

Visiting Roxboro after eighteen months of intense training, Ned saw no young men left. Growing-up here helped instill his lifelong work ethic. Add the discipline of being a pilot, the skills and judgment he acquired, all reinforced the challenge to give his best. Long forgotten was his father's warning when he'd been a young boy fascinated by airplanes.

"Now, Nooks, don't you dare think about going up and flying one of those crazy things."

* * *

Hometown friend John O'Brian drove Ned to the rail station at Danville, Virginia, and saw him off to Texas. So numerous were servicemen who passed through Danville's depot then, that townspeople used to meet the trains, eagerly hoping to swap a cigarette or other ration they didn't want, and get a chit for sugar, flour, or coffee.

Debarking in San Antonio, Ned became part of an aspect of U.S. aviation training unique to World War II. Young pilots taught each other. CIS or Central Instructors School at Randolph Field, pushed top graduates through an intensive course that prepared them to instruct cadets coming behind them. Experienced older pilots did not exist in sufficient numbers, a dearth that prevailed through the war. Civilians still taught primary flight in schools like the one at Douglas, Georgia, but only military pilots instructed at higher levels. In practical terms, Ned soloed in a single-engine fabric biplane in January, 1944. Nine months later, he'd be giving other aviation cadets advanced instruction.

The program taught rated pilots to master the process involved in learning how to fly, to understand and empathize, to be back in the shoes of a training cadet. The twin-engine aircraft they flew defy modern description—the Beechcraft AT-10 Wichita, or its close kindred the UC-78 Bobcat, a military version of the Cessna T-50, nicknamed the "Bamboo Bomber," not for jungle valor, but expedients of construction. Both planes were built to the extent possible with non-strategic materials, by ingenuous use of metal skeletons framed up in wood or plywood. In these low-wing cantilever machines (they didn't have the exterior wing supports of biplanes) instructors across the nation took their fellows into the air. They shifted from the left seat back to the right, where they'd sit alongside the cadets assigned to them. They attended classes on teaching effectively. In the cockpit, they felt again what it was like to be a student. They accustomed themselves to using the left hand for the throttle and the right on the yoke till every move became second nature. Here were the men who either helped other pilots succeed or dealt forthrightly with those who could not. Ned flew five days a week, welcoming the relief of powering his A-10 up to 5000 or 6000ft to escape the furnace of the Texas heat, and let cool air through the vents in the manifold.

Saturday dinner at the officers' mess was a treat, with tamales, tacos, enchiladas and other Border specialties. Such dishes are national favorites now but at the time, Ned and many others had never tasted them. CIS also offered the novelty of a couple of weekends for sightseeing in San Antonio. Ned remembered the beguiling western flavor of this medium-sized city drenched in Texas history. He and a classmate struck up for a tourist outing on his twenty-first birthday.

"He'd been a state cop in Rhode Island, so he was a few years older. We're out seeing the Alamo and some of the sights, and he turns to me and says, 'Let's go get us a couple boiler-makers.' I didn't know what a boiler-maker was, maybe something they built in the factory neighborhoods up north where he came along, but I wasn't going to show my ignorance. So we step into a bar, he orders, and what he's talking about is a beer with a shot of whiskey. It was sort of a coming of age for me. He chugged his down, I barely sipped. What a waste of liquid."

On September 25, Ned reached Shaw Field in Sumter, South

Carolina. The 60-day course he taught was equivalent to the Advanced level he completed at Moody, with the name changed to Basic in a reconfigured program. Each instructor was assigned several cadets. In late life Ned remembered bringing them along, taking them into formation flying, and that they all passed.

His next orders were cut to the 815th Army Air Force Base Unit at Malden, Missouri, for operational training in the C-47. An administrative delay gave him an unexpected week to return to Roxboro, and he and a Moody classmate he'd be seeing more of, Jack Mayer, from Rock Hill, South Carolina, got places on a private four-seater plane to Atlanta, where Ned figured he'd catch the train north to Durham, then hitchhike home. Hoping for news of his college roommate Deming Ward, he phoned the Wards in Durham, and unexpectedly found his mother there. Here he also learned his friend had gone to the Orient. The sudden reality of this land of mountains, its "forgotten war" and the remote, unearthly-sounding place name of the Hump, were closing in. Ned himself would have faced imminent deployment had it not been for delays at CIS and instructing at Sumter. Might he and Demming be meeting up again before long?

He and his mother used a precious gas ration to drive out to the tidy country home of the beloved cook who'd faithfully worked for the family through Ned's childhood. Good-hearted Rebecca ran across the porch, a beaming smile splitting her face, and threw her arms around him, thrilled to see the boy she'd had a sort of proprietary interest in raising, now a man in a uniform with wings on his chest, and she could not admire the insignia enough, simple as it was.

A few days later he was in Danville again, to catch the train to Malden.

CHAPTER 3

"PRIVATE PROPERTY: NED THOMAS"

MALDEN ARMY AIR FIELD HAD FUNCTIONED AS A PRIMARY FLIGHT facility from 1942 until July, 1944, when the base was commandeered into operational training for Troop Carrier Command pilots. Hailing from around the country, they'd flown a variety of aircraft; a handful had been instructors like Ned Thomas; but whatever their backgrounds, all found their skills urgently required now for transport. Each probably wondered, too, whether he'd be sent into the waning gasp of war in Europe, or out to the inferno of the Orient, where hostilities still blazed with no end in sight.

Sprawling as a desert, bestrewn with hangars, base living, classrooms, administrative and support buildings, Malden Field did not abound in home comforts, and the place seem situated mainly to funnel in every wind that blew. Single storey BOQs, or bachelor officers' quarters, had each four unheated bedrooms with a pair of double-bunk beds, and a central gathering room balmed and sooted by a pot-belly coal stove. Latrine facilities required a chilling walk to a separate building, as did getting to the mess hall. The wintery weather already held a penetrating iciness, nowhere more than along the expanse of the flight line, where hard winds tunneled down with numbing force. Pilots looked forward to getting into the cockpit in the morning just to start their heaters.

On the plus side, many unattached young adults worked at Malden, and the place offered a livelier social scene than cadet life. The town boasted a restaurant or two and a movie house if a fellow could borrow a car. But the base with its opportunities to gather and hobnob really functioned as their social hub —its movie house, and in an era of big bands where so much socializing was done, the officers' club booked a combo that kept the dance floor filled.

The flight line functioned as a major informal social link and was one of the most popular spots on base. Pilots encountered friends, matched up at the ping-pong table, or strolled in with their dates. The snack bar stocked sandwiches, cold milk, hot coffee and desserts. A radio at full volume could barely be heard over the roar of the planes flying in and out, all hours, day and night. Given the electricity of constant comings, goings, and greetings, the place projected the outgoing nature of wartime Malden in capital letters. Newcomers were befriended. Everyone braced up to the constant goodbyes. Who'd have wanted to risk losing that once-in-a-lifetime encounter or introduction?

It seems inevitable that a handsome pilot and Malden's prettiest lady would meet.

Neva Rae Hudson, a slender blue-eyed brunette with the face of a movie star and an outward vivacity that somewhat covered a bit of old fashioned shyness, worked at the airfield administration keeping track of fuel used on base and in aircraft. Raised in neighboring Arkansas, the only daughter of a lawyer who'd been a city attorney and state congressman, when her parents relocated to the Missouri town of East Prairie to look after some family farms, she'd taken a job at Malden Field, where she shared a minuscule base apartment with her close chum and coworker Lou Marshall.

The couple's introduction was engineered by a fellow pilot who borrowed a car into which six or seven of them scrambled, and they set out for the town bakery to sample the sweets that got pulled from the oven every Saturday evening, at a late hour. Being a night Ned and Neva Rae both remembered with great mutual lucidity, perhaps only their exact words can convey it.

"I met her on a blind date."

_____ "I met him eating doughnuts."

"She was working there at the base in administration. Doing something with fuel."

_____ "I broke another date to come."

"With one of my instructors."

_____ "I believe Ned had been dating this pretty widow with young —"

"Oh Neva Rae, that was nothing serious. "

_____ "I broke a date to come. First time I'd ever done that. I said I had to come home early, I was doing something with Lou, which I was. She came, too."

"This fellow borrowed a car, I think from his father who lived over in Mississippi."

_____ "He was a good friend of Lou's, but they weren't romantic. He was always so funny. Do you remember him?"

"Sure. Loaned me the car a couple times for us to go out. Nice guy."

_____ "We went to the bakery late Saturday night, when they took the doughnuts out. Everyone was laughing. Crowds used to come there from all over town on Saturday evenings. We were just being silly as kids, everyone was. I don't think I ever was so silly."

"They were pulling out these doughnuts and loaves of bread, lined 'em up at the counter."

_____ "It wasn't just doughnuts, they had cakes and pies and all these little fancy goodies. People were standing around eating, some of them a lot older than we were. The later it got, the more people laughed and talked."

"What was the name of the guy with the car anyway? Jack something. It wasn't Jack Mayer. Another Jack."

_____ "They brought Lou and me home late because the baked things didn't come from the oven till midnight."

"They took everything from the oven around eleven."

_____ "Ned saw me to the door. He kissed me."

"Oh, Neva Rae. . . ."

_____ "Only time in my whole life I ever let a fellow kiss me on a first date."

Next day, the friend with the car barged into administration, faced Neva Rae at her desk, and asked point-blank what she'd thought of Ned Thomas. Gazing up from her slew of papers and ciphers, she refocused two jewel-like azure eyes.

"Well, I guess he's cute," she answered discreetly.

"Cute," he barked. "That's what you call tall, dark, and handsome? Sheesse."

Cute or handsome, the two made a stunning couple. Every day after flying, Ned walked over to Lou and Neva Rae's. They'd head for a movie or the Officers' Club or the flight line. They strolled along bundled against the freezing cold of Malden's winds, holding hands and talking. There was her famous dinner party of raw chicken. She was not yet a cook, though she became a great one. Ned and Lou managed to saw up the business at the table and reheat it decently. At Christmas, she headed home to East Prairie. Pilots had no leave for the holiday.

Early in January of 1945, having to report for updated regulation dog tags, Ned asked the technician to bang out an extra one with a special message. That evening he handed Neva Rae her own dog tag engraved with the words: "Private property. Ned Thomas."

* * *

If anyone was looking hard at wartime airplanes, it was pilots whose lives depended on exploiting their strengths or compensating for their drawbacks —of the latter, always more than a few. Virtually all major World War II aircraft were launched under inadequate testing for real-life conditions. Take the Mitchell twin-engine B-25 Jimmy Doolittle flew on his raid to Japan. The plane lacked autopilot, cabin pressurization and turbos, leaving Doolittle and his crews to maul their way across every mile of ocean they flew. Whatever came of the mission, it stands a tribute to great pilot work, especially considering the modifications and adjustments jurry-rigged on site virtually up to takeoff.

The bigger B-17 Flying Fortress had modern turbos, autopilot, and longer legs, but the thing guzzled fuel like a drunk, and had a way of sprawling helpless on the tarmac as mechanics tried to get it back on its feet, so to speak, and flying again. Ned was intrigued by a newer four-engine he eventually flew, the C-54, an adaptation of the commercial Douglas DC-4 Skymaster, sort of a "bigger brother" to the C-47 he came to Malden to fly. With its extended range and capacity, the '54 gained favor with generals, and also with men who aspired to be commercial pilots post-war in the best of up-and-coming airliners.

But bigger wasn't always better, not for military use. The C-54 was too large to tow the gliders that moved troops into battle positions. In the Orient, its low ceiling made it unsuitable to murderous Hump routes, particularly the higher upper range. (It will be remembered the southerly route did not reopen until Burma's liberation). This left the C-47, though it, too, proved seriously deficient without the turbos needed to perform reliably at high altitudes. In fact, the thing killed crews past number. If nothing else, the brutal flying conditions of the Himalayas forced the military to confront complex life-or-death tradeoffs among various aircraft.

In the purview of engineers were the strengths and deficits of the reciprocating engine versus the newer jet aircraft in the testing stage. Plainly put, the reciprocating engine was less reliable, strong reason to respect pilots who jockeyed these temperamental wing-and-a-prayer machines around the world and into wartime missions. "Propellor aircraft had so many more parts to rattle around and go wrong," Ned says bluntly. They couldn't fly as high as jets, or lift above weather patterns, artillery, or other hazards. Some were air-cooled, some liquid cooled, though most, including the C-47, were air-cooled. The fly in the deal was that while an antifreeze component was more modern, if an engine took a bullet, pilots found themselves out of luck, whereas good-old pure air was there for the older craft, provided they weren't cruising too high for oxygen. One of the most positive developments in war-worthy aircraft was the sudden interest in improved transport technology, a challenge never before seriously addressed. Ned credited these advances for saving the lives of many pilots who in the older converted bombers would not have lived out the war.

In sum, the C-47 was a workhorse, but small with other drawbacks. The C-54 lacked altitude. As for the bombers —no matter how converted or reconfigured—they remained volatile and prone to killing mishaps and snafues for troops and transport work, as the Hump bore solemn witness. It happened, too, that Uncle Sam wasn't deploying bombing and fighter missions in great number now. More pressing was the need to move materiel and troops.

Did Ned, like many pilots, think of carrying his training into lucrative civilian aviation? He was good at flying, and what made planes operate intrigued him. But he also had something of his father's entrepreneurship. As a personable lieutenant of twenty-one, he did not imagine that flying would remain in his spirit. That having soloed in a depression-era canvas biplane he'd eventually go up in the F-4 jet. "If it had a propellor, I've probably flown it," he joked long after. Yet somewhere between the years of the big noisy planes he flew, and his developing people skills, his personal qualities would coalesce uniquely with his keen knowledge of flight to take him through an exceptional life.

Meanwhile he'd come to Malden for the C-47, the most versatile and widely used transport to that time, the pack mule and ferryboat of war. Two jeeps or three aircraft engines could be loaded through the widened cargo doors. Fully loaded, the plane hauled 28 troops with full armament, or 15,000 pounds of cargo weight on the floor of its strengthened fuselage. A glimpse along Malden's flight line gave a wide-angle view as to why the '47 was called the Gooney Bird, for its distinctive flattened beak. Adapted from another Douglas passenger aircraft, the DC-3 Dakota, it was said by some to be the first to fly the Hump, though this probably is incorrect. Rugged, capable of attaining a reasonable 230 mph in ideal flying, the drawback of the Hamilton Standard three-blade hydraulic props was that if an engine stalled, in order to stop the prop from windmilling the pilot had to "feather" fast, that is, maneuver the blades into a position vertical to the wind like a knife edge instead of the flat edge that can create deadly drag on a stressed aircraft. Windmilling was one of the worst bugs in Gooney's beak.

A real lynchpin of the war, the C-47 carried uncountable soldiers to the battlefront for parachute drops, or towed their gliders. Pilots superbly mustered this comparatively clumsy plane into hostile territory at low elevation to avoid enemy radar, and still in formation, rapidly climbed high enough for paratroopers to jump to their engagement sites or for an attached craft to disconnect. Low flying reduced radar risks but made planes extremely vulnerable to ground fire, so timing had to be split-second in precision. Fly under, pull up, deploy your men safely, gun your ship away from the line of fire, all without colliding into the fleet of other aircraft at your shoulders. One recent contingent of Malden pilots had trained for the C-47s used in the Battle of the Bulge three months earlier, in August, 1944.

This vital glider training took a lot of their time, as Ned describes. "First we rode in the glider with the glider pilot. That's so we can see what he's experiencing when we pull him, and watch him release and detach the glider and glide in. A glider could carry maybe 15 or 20 troops with equipment and weapons, or if there was extra equipment, we might have to displace some personnel. Gliders were canvas and plywood, had some metal struts and little wheels. They were used a lot on D-Day. Training in the C-47, we had to circle back and drop the towing rope, too, but that isn't done in combat conditions. We probably were towing about 120 miles per hour."

It was precision, journeyman work.

* * *

At the ebb of 1944, victory across the Atlantic mainly promised the grueling mop-up of a war-weary continent. The Normandy invasion on June 6 was a success. In August, Paris was liberated. In September, Americans landed behind German lines at Arnheim in the Netherlands. December brought the Battle of the Bulge, and the cutting off of the German counteroffensive. At last, the Allies proved unstoppable. In January, 1945, Franklin Delano Roosevelt, ill and near death, recited the oath of the presidency for his fourth term in office. Two weeks later he appeared at Yalta beside Churchill and Stalin, to parley Germany's coming defeat and occupation, an outcome not yet achieved, but which few doubted anymore.

It was another story across the immense sweep of the Pacific, where matters hardly could have looked more bleak for America, despite news of victories. In June, forces under Admiral Chester Nimitz fought a month-long battle to retake vital Saipan, whose air strips later launched the atomic bombing raids on Japan. The Japanese sustained heavy air losses in the Battle of the Philippine Sea, and in July, Americans landed on Guam. While Ned had put air cadets through their paces at Sumter, General MacArthur's troops began the invasion of the Philippine Islands, where months of agonizing ground fighting, from October into early February, would bring them at last to the capital at Manila. There seemed no end. That February, U.S. forces would begin to fight at Iwo Jima, where on the 23rd, U.S. Marines raised the flag on Mount Suribachi, though heavy combat would continue there to late March. Horrors yet loomed on Okinawa and other scattered battle stations. Weary GIs saw nothing ahead but an intractable series of sea encounters and island wars against an enemy who preferred death to surrender. For U.S. pilots, it was a pretty sure bet that the most pressing needs would exist henceforth in Asia.

If the Chinese were in a losing conflict (and General Chennault continued to argue they could win with sufficient aircraft) hopes persisted that by supporting and supplying Chiang Kai-shek, the enemy would stay hobbled inside his country's borders. Had Japan chosen to pull sizable contingents of this occupying soldiery into the Pacific against the U.S., the carnage for American servicemen fighting their way across would be past terrible to contemplate, horrific as it was. China's vastness, primitive conditions and undeveloped infrastructure in a territory 300,000 square miles in excess of the continental U.S., her pathless mountains and interior swaths not yet penetrated by roads usable for military egress, all combined to create a landscape tricky and evasive to permanent occupation or conquest. If Chiang only could hold on, Japan finally might find herself too mired down to fight effectively.

Imperial Japan did have a stellar advantage in China's sky. Claire Chennault assessed the fact astutely. Given the ability to harry from the air, their commitment of such an immense troop presence for an endless ground-pounding occupation seems anachronistic, even baffling, though in the cut-throat wars of the '30s, the barbarous tactics had proved effective, and succeeded, too, in keeping China's coastline tight in the noose. So having these hordes of Japanese fighting men bogged down in China was a situation by which the U.S. had much to gain, at least for as long as American GIs were looking down the sights of their guns in the Pacific.

But war is unpredictable and complicated. Whenever

Washington expected miracles from Chiang Kai-shek, disappointment followed. Nor could Uncle Sam spare extra servicemen to his cause, not given the multiple fronts on which the U.S. was fighting. In countless ways it devolved on the men flying the Hump to underwrite the price of their country's requirement for China's struggle. History testifies that these unsung pilots accomplished far above what they were sent to do. The Hump supplied the Nationalists, as well as several U.S. detachments in Asia, and fueled a war cut off from all lines of supply overland.

Of several U.S. commanders who played critical roles in the CBI Theater, Chief of Staff General George C. Marshall must come first. Fed up with Chiang's tepid fighting and the appalling fatality rates over the Hump, Marshall wrote the President and Joint Chiefs blisteringly. End all support to China and pull out of the Hump. Marshall's words carried a lot of authority, but the downside could not be ignored. Abandoning China would strengthen Japan's hand and shore up their resolve to keep fighting. Marshall also allied himself with Stilwell in proposing that Vinegar Joe be made leader of China's fighting forces, and Chiang dealt out. Roosevelt overruled him, recalled Stilwell, and replaced him with Major General Wedemeyer, who as the new commander in China (though not the entirety of CBI) proved more effective at diplomacy with the Generalissimo, and at addressing problems of training and equipping his men. With Wedemeyer at the helm, Operation Grubworm commenced, and the U.S. Tenth Air Force began transporting divisions of Chinese fighting men, as well as flying the notorious 24-hour Hump schedule that would remain till war's end.

In a mountainous region where radio and navigational capabilities were little-to-nil, where means to illuminate airfields after dark a luxury and rarity, the introduction of night flying over the Hump marked a leap in tonnage, but an exponential spike in fatalities. Pilots gritted their teeth, commemorated their dead whom they could not recover, and grew cynical. Wedemeyer can be credited with enhancing the Hump's potential. But only provided his men could live through it. Another big if.

Under the subsequent leadership of General Tom Hardin, the Airlift expanded again, in large part due to his hated order, surely the most infamous ever issued in the theater, "Effective immediately, there will be no more weather over the Hump." This meant no coddling of pilots, as if they ever flew an easy mile. No scrubbed flights, no matter how many crews went down, no matter what conditions prevailed. "Tom was a driver," writes his successor, Brigadier William Tunner. "He drove himself and he drove his men."13   How hard did he drive?   By trying to erase the Himalayan weather challenge, the already horrific loss of human life soared over the summits. But if tonnage was all that counted, Hardin delivered his pyrrhic victory, if not something close to a mind-numbing military wonder.   Still the truth remained, and no one could gainsay it.   There was no good weather over the Hump, nor even marginally good.   This, and the limits of aircraft technology, were the constants.

General William Tunner, the last commander of the Hump, revoked the despised order and oversaw some reduction in fatalities.   Improved pilot training began to factor in.   But like his predecessors Tunner drove his bargains.   To demands for more flights, he countered for more airfields, planes, personnel, navigational aids primitive as they were, and ground crews.   The best thing General Tunner did was to rethink the concept of airlift from top to bottom, and make transport flying a science, if not an art.

He writes, "A maxim proved there on the Hump was that only such men schooled in and dedicated to air transport, can direct this complex new military service with full efficiency.   In the adversities, the extremes, and the eventual triumph of the Hump Airlift was cast the mold for all future air transport, both civilian and military.   Airlift proved itself not merely feasible, but practical, and superior to other transport in many ways.   The most efficient methods were developed there in that laboratory at the end of the world. . . From the Hump on, airlift was an important factor in war, in industry, in life."[14]

Concepts like proper loading and balancing of aircraft and meticulous maintenance, seem obvious now, but planes were flying the Himalayas without scheduled service, out of balance, their payloads "just stuffed in any place." Pilots, too, needed rotation on the roster for rest between flights. Tunner demanded men with better training and more experience, and he was one of the first to recognize that transport had a different rationale from any other type of flying.

"Their very mission was different. A combat pilot is to inflict damage upon the enemy. If he doesn't, well, you expect losses. The mission of our pilots was to move their planes toward their ultimate destination . . . to fly skillfully and safely. Our pilots flew over the roof of the world, as the Himalayas are called. They flew over vast deserts as well as over the oceans not only under constantly varying conditions, over widely different sections of the earth's surface, but different types of planes as well. He might suddenly find himself making a landing at night, with two Zeros on his tail, on a Chinese landing strip. He might fly two thousand miles, king of the cockpit, master of the ship, then come home as just another tired passenger in a canvas seat on a slow but reliable C-47, tossing and pitching for hour on hour through the sky. Our pilot was by no means a drudge; he could wear his hat as sloppy as any fighter pilot. They got into, and out of, scrapes in dives and districts all over the world. They were not angels. But they did know how to fly airplanes, and fly them safely and securely."[15]

The history of the Hump with its countless dead, quashed for all time the notion that vital transport could be done piecemeal, or by distracted field officers in other commands.

"I found out there was a lot more to flying than simply boring holes through the sky," General Tunner notes with pride.

"Neither planes nor men could fly constantly. Planes required maintenance, men required rest . . . We proved that air transport is a science in itself; to be carried out at its maximum efficiency air transport must be run by men who know the techniques, and who are dedicated—professionals!" 16

After the war, Tunner took his CBI experience to Europe to manage the Berlin airlift. From Tokyo in 1949, he commanded the operation for Korea. Time brought recognition of the Hump as the great prototype for this critical lever in the machine of war. Airlift was taking off.

* * *

March 1,1945, brought Ned's orders to report to the 807th Army Air Force Unit at Bergstrom Field, Austin, Texas, for four weeks on a larger aircraft, the C-46 Curtiss Commando.

At Malden he and Neva Rae had seen each other every day. He'd land his plane, phone from the flight line, they'd be together when she left her office. They were in love and things had happened fast. No time for rings or a party. They pledged to write each day, and their quiet engagement would conclude with marriage, as soon as he returned from the war.

On the evening of March 7, they said goodbye. In the gathering dark he headed back to quarters to get his bag and join the other men boarding the troop carrier train.

Neva Rae dashed through the evening chill with a handful of files for her boss, who lived three doors down, one of the few places on the street with a telephone, and he'd been generous about letting the roommates receive phone calls there. She'd encounter other company, and right now she didn't need to be idle or alone. She couldn't have felt very happy.

Striding rapidly along through the heavy wind and thick shadows, Ned stopped in his tracks as a new idea thrust its novel anatomy to the front of his mind. Should he and Neva Rae really wait, or might they have another choice? Momentarily immobilized, he tugged at his thoughts and weighed things. The train would start boarding in an hour. Orders had him due at Bergstrom the morning of the 10th, three days off. So long as he reported in on schedule, he was free to manage the interim as he saw fit. But what kind of marriage could he offer in the chaos of preparing to deploy? Wouldn't it be better to wait for peacetime, when he got home again?

Would he get home? Could he wait? A pilot learns that split-second decisions can spell life or death, though maybe now, happiness or heartache. No time to hang back second-guessing.

Neva Rae had seated herself and opened her paperwork when a decisive knock sounded at the door. Rising in wonder, she gazed outside at Ned standing on the narrow stoop, a taxicab idling at the curb behind him. His words were all she needed to hear.

"Let's go to Texas."

"He came back," she marveled. "He wouldn't let me say good-bye, after all! And I decided right there I'd go with him. Right in front of my boss. I just quit on the spot and left my job. I walked down three doors, packed a bag, left a note for Lou, and tried to think what I'd need, while the taxi driver sat with his engine running."

Women weren't permitted on troop carrier trains, so Ned directed the taxi to the depot in Dexter, fifteen miles north. Around midnight they boarded a train bound for Oklahoma City. Numb with joy and hope, a dash of trepidation, the certainty of love, they spent two days and nights waiting at stations and sitting up on the wooden benches of the railcars, which refilled and burst to the aisles, scarcely space on the hard seats to accommodate them, no place to lay a head. Early on the morning of March 9, they pulled into Austin, Texas, disembarked, and according to plans hatched on the journey, perused a newspaper over breakfast at a café nearby. From a phone booth they made a few calls, then headed to City Hall, where Neva Rae sat outside on the steps while Ned went in and secured their marriage license.

In a listing they located a room in the snug bungalow of a young college teacher and his wife, then they rushed to the Central Methodist Parsonage of Austin, where they were married in the living room of the Rev. Frank D. Charlton, whose wife served as witness. Neva Rae wore the plaid wool suit in which she'd spent two days on the train. Ned stood tall in uniform. Following the modest ceremony, the bride sent a wire to East Prairie, to tell her mother and father of her elopement with the wonderful pilot she'd been telling them about. Ned posted a letter explaining things in a general way to Miss Rossie of Roxboro. With matters carefully looked after, they went to their first dinner as man and wife. Gazing into each others' eyes, they pronounced the evening wonderful, then headed hand in hand back to the professor's.

* * *

Across the flat windy expanse of Bergstrom Field, in the white glare of a hot Texas sun, the C-46 Commando loomed like a dinosaur taking its ease on a tripod of legs. From a distance the great sweep of its wings stretched into Ned's view, then the huge front wheels that hugged the tarmac as if to mime the immense forepaws of a sphinx. The rear tire strut appeared bent at kneecap to balance the ponderous gut of the fuselage and the proud outsized flag of the tail, for like a brontosaurus the plane was a tail-dragger.

At the wheel of a staff jeep, a corporal steered into the beast's range at low speed, made a smooth turn under the wide chest of the fore section, swung into the shadow of the outstretched left wing, and pulled up at the rearward door. Alighting from the shotgun side, Ned stood for a moment in the gusty blow of wind and dust, and looked on in silence. Then slowly he began walking the circumference of the monster.

In the sky this Goliath could snort fire and brimstone. Ned's job was to face and tame it, and by no stretch did he see here the tentative look of a half-taught colt. He fanned his gaze upward at the wide unblinking Cyclops eye of the cockpit, the big blunted nose that could bore a hole through a polar ice front, props that revved like whirling claws against the wind, each of the four blades taller than Sinbad and deadlier than his swinging sword.

This big shot had a reputation.

In one of the more startling photo-ops of aviation, a Curtiss-Wright C-46 rests on the runway with a hundred men standing shoulder to shoulder across the top of its 108-foot wingspan. Nose to tail, the 76-foot length exceeded that of the sprawling four-engine Liberator. Muscular rather than sleek, the mass of the aircraft assaulted the eye like a bull-grizzly wheeling to charge. Turbos in the two enormous engines set the aircraft's ceiling at a robust 26,000 feet, but it would climb higher when pushed. The thing cruised at 264mph in fair weather, and lifted some 20,000 pounds of cargo if regulations were adhered to, more if they weren't.

Notorious for its uneven temperament, the Flying Coffin was a byword. It was no laughing matter to put a flyer unschooled in its ways at these controls. Recent months had brought improvements and boosted the plane's repute to a degree, but pilots still had cause to view it warily. Like the Angel of Death, the thing took a lot of men to their graves.

Ned, his friend from cadet days Jack Mayer, and another pilot who finished high at Malden, had got waylaid en route to war for final operational training on this, the biggest dual-engine transport on earth. The strengths of the plane had been latently recognized, the drawbacks addressed, the deadly lessons pondered. But it took know-how to fly, even for experienced pilots. On a personal note, Ned could thank the detour to Bergstrom for making possible his elopement with Neva Rae, though their time together would be curtailed. He'd be flying nights a lot. Night flying meant instruments, and by now his future loomed to the westward, for island transport, ocean or mountain flying, maybe troop work. In a territory comprising half the globe, he'd be dependent as never before on his dials. Nor did it take long before he was making comparisons between the C-47 he'd flown at Malden, and the bigger C-46.

"It had problems to begin with, but I preferred it. Compared to the '47, it had more capability, it was faster and had more innovation to it. At the time, it was state of the art. It seems primitive in retrospect. Rough too. They didn't have power controls back then, and the C-46, honey, you had to manhandle it. Later they tried putting weights out on the ailerons to make them a little easier to move, but that wasn't till after I got back from China.

"It was a good plane really, and turned out to be a heavy hitter. It had a lot of problems initially, but by the time I got to the '46, I'd been an instructor, flown the '47, and the kinks had been worked out pretty well. One thing the old '47 had was flexible wings. I've watched them out there flapping in the turbulence. On the '46, they were built tough, though maybe the '47 gave you an easier ride."

In retrospect, of course, flying the C-46 was quite different from the aircraft of modern time. "These planes were tail-draggers, remember. Most pilots nowadays haven't experienced tail-wheel planes. Say you're taking off. The tail comes up, that wheel's not on the ground now, but you still want a little nose-up-tail-down, for lift. Because if the nose is down, you're losing lift. So you push the stick forward to get the tail wheel up, and get rudder control of the aircraft. You're approaching flying speed, your tail wheel's still on the ground, and the winds are coming from different directions, so to control the plane you have to use the brakes till you can lift."

It sounds almost like the pilot had to slow down just to speed up. "You're going down the runway. You advance your throttles, release the brakes and start rolling, gaining speed till the tail comes up, which you need to get rudder control. Otherwise you're just using your brakes. Now, you're moving into position, advancing throttles to takeoff power, using the rudders to steer. When you reach flying speed and break ground, you pull it off pretty fast.

"You start to ascend, give your copilot the thumb signal, and he brings up the gear with the gear handle. The idea is to clean the plane up. Get all your garbage in. That's in case you lose an engine on takeoff, so you don't have the gears or flaps creating drag. You don't want the wheels still down, there's not enough runway length. You need everything up, till you can turn the plane around, reapproach, and put it back down."

Life and death were in every detail. The rudders: "They control the right and left movements of the plane. Your pedals also contain the braking system. The rudder pedals are below the dash, with the brakes at the top. For the rudders you use the heels of your boot, for brakes you raise your feet upward on the rudder pedals and depress them, either one or both together. If you turn the plane, you're going to use the throttles, the left to turn right, or the right to turn left. That's true on a nose-wheel plane, too, though nowadays the nose wheel is steered. With a tail-wheel plane you've got to use braking and opposite power.

"Say you've got a crosswind, you have to brake some to keep a center line. As soon as you get rudder control, you drop your foot down and use the rudder. Now the tail of the plane is called the vertical stabilizer, the back wings the horizontal stabilizer. The rudder is along that vertical edge of the tail. And your flaps are at the edge of the front wings. You hit that crosswind taking off, you adjust your right and left rudder pedals, because when the wind hits the plane, that stabilizer is just a big wind catcher. You're taxiing out, trying to gain speed to fly, and if the wind pushes the tail to the right, the nose gets pushed to the left, or vice versa."

And for landing? "A plane's just a leaf up there in the air, with nothing to hold it in any direction. Say you're on final, the tricky part of landing is to be sure the plane is in a definite direction to hit the runway. Unless you do something to compensate, the wind's going to push you off course, so you adjust by crabbing. That means turning your plane into the wind. When you touch down, you can't be at an angle or you'll sheer the landing gear off. So either you time it where you can kick it out of that crab and put it on the runway before the wind can blow you off course again, or else you lower a wing, and that gives you enough stability to come down. The only problem there, you land on one wheel. So you may use both, lowering the wing, crabbing, you jockey it in." For a pilot it was all in a day's work.

* * *

What wasn't on the work sheet came with an unexpected peal on the professor's telephone. From the end of the long-distance line the enthusiastic voice of Ned's mother announced she was on her way for a visit, and would arrive at the city airport next morning.

"I can't wait to put my arms around you!" she poured into Neva Rae's startled ear.

They'd been married two days. Time for Mama to check out baby's new bride.

For once Ned looked helpless. "Think you could go meet her? I'm on the roster to fly."

Next morning, Neva Rae watched a tall woman with a resolute face descend the gangway. Planting a brave smile, keeping the small talk flowing—a skill of which southern women then excelled verily from the hour of birth—Neva Rae hailed them a taxicab, conscious of a somewhat heavy sensation. Ned had scouted out a nearby room for his mother but she preferred to wait at her son's. Besides having Ned in common, both women had taught primary school, and in the course of things, Miss Rossie would have gleaned a bit of pertinent data.

Born in Atkins, Arkansas, in 1921, Neva Rae Hudson had been raised by her soft-spoken mother Mary, who despite a reticent manner, ran the bank during the Depression, stitched superb couture, and loved dressing her daughter up. Her father, Raymond Samuel Hudson, who listed his occupation as "farmer" on her birth certificate, was a lawyer and state legislator who turned down the chance to be Attorney General of Arkansas, to make a run for the U.S. Senate, which he lost. Their daughter attended a select female college in Mississippi called Blue Mountain, noted for chaperoning very ladylike young women. Some of the regimen may have grown confining in the view of Neva Rae, who though liked by her teachers and peers, transferred to co-ed Arkansas State. In her third year she left to accept a teaching position at a school in Risco, Missouri. Country pupils fell head over heels with lively Miss Hudson; one little fellow poured in wobbling letters: "Dear Teacher, you air lovely. You air dancey."

That New Year as a passenger in a car on an icy road, she was struck from the rear by a truck and thrown from the back all the way through the front window. She spent weeks in the hospital, where she made up her mind to show her patriotism and join the Women's Army Corps. It didn't hurt that she was a crack shot, having target practiced in the country as a girl. Mannerly and poised, she impressed commanders who sent her traveling roundabout Idaho and Michigan to acquire know-how like spotting and identifying aircraft, and standing up in a turret to sight and shoot. They quickly determined to send her to officers' training school, but when her enlistment ended, so much did she want to return home, she protested that she was thinking about getting married.

She may have been considering a proposal or two. Men looked at her and lost their heads. Her dance cards filled before she set foot on the dance floor. Reporters at Malden Field's newspaper dogged her steps to snap her picture till she begged them to stop. To think what a poster girl she'd have made for the WACs! When she met Ned, she assumed he was a few years older and was surprised when he told her his age, though of course, it hadn't mattered a bit. How much all this registered with Miss Rossie is anyone's guess. Her eagle eye did take in a burn on her daughter-in-law's forearm, and she softened when Neva Rae confessed she'd got it trying (again) to prepare her husband's favorite dish, fried chicken.

"I think after that we were all right. I used to get up before daylight and send Ned off with a good breakfast, then do our laundry in the bathtub and hang it out back on the fence. His mother was there every single morning at eight. She must've stayed for weeks."

"A few days, Neva Rae."

"It seemed like weeks."

Rossie Thomas cast her piercing eye on the hands of the couple, and wooshed them to a jeweler, so perhaps to her martinet of a mother-in-law did Neva Rae owe some diamonds that adorned her ring finger from that time.

Speaking in later life of this long, enduring marriage, Ned names the single worst and single best decisions he ever made. "The worst decision I ever made was to marry Neva Rae. It wasn't fair to her. I was too young and immature. I had no prospects or plans. No stability to offer, nothing settled in my life or work. But about the very best decision I ever made. That was to marry Neva Rae. With her I couldn't afford to tarry around and put things off. I had to grow up fast, and take responsibility fast, and it was the best thing that ever happened to me."

On April 14, he boarded the troop train to Baer Field, at Fort Wayne, Indiana. A major point of deployment for pilots, it was at Baer that the men's records and inoculations got a final update, and they passed the pseudo-physical that was a farce at best. "We lined up, got the tongue depressor, they looked in our ear, and passed us. We were pilots, we were trained, we were heading out. Only thing gonna stop us, was if we had a leg missing."

Neva Rae had agreed to drive the pregnant wife of another pilot to Oklahoma City. Writing off her own nausea as a case of flu, at their destination she convalesced for a few days with a great-aunt whose income was enhanced by an active oil derrick behind her big elegant house. Then on the spur of the moment, she decided to catch the train to Fort Wayne. Here she contacted a nurse friend who'd found work in the town, and helped her locate a room near the base. As if rehearsing for a spy film, she sent a telegram to Lieutenant Ned Thomas at headquarters, containing a cryptic message. His "radio had arrived" at a certain address.

He came the minute he could, but it was a last parting. He'd been assigned a new C-46 just off production at Curtiss, and was standing by to deploy.

Next morning she wakened to the jangle of the phone at her head.

"Look out the window," rang her friend's voice. "They'll be coming over!"

Neva Rae pulled up the sash and leaned out over the sill. A rumble like gathering thunder began to fill her ears and shake the house. The planes were flying out.

Wedge on wedge, wing on wing, they came, flinging their broad shadows against the ground, their "helluva roar" blotting out earthly sound—on their way now to war.

CHAPTER 4

FLYING THE FIREBALL

SLEEK OFF THE LINE, ITS ALUMINUM SKIN GLEAMING LIKE A new-minted silver dollar, the C-46 spangled back the glare of the morning sun. Within, a breath of scoured freshness wafted from the cavernous bay, whose pristine depths not a mote bestirred. Presently housing a new aircraft engine in a massive crate, the impressive space was configured to accommodate sixty troops on canvas benches along the walls, or thirty-three hospital litters and their attendants, or three jeeps and a load of cargo. Even the cockpit was buff in a scrub of soap and wax.

Up at the controls, Ned Thomas and Jack Mayer covered their tasks without technical commentary. Jack at present occupied the left seat for which they'd drawn matchsticks before take-off. Nonessential for now and sprawled on the desk at the cabin wall behind the bulkhead, slept a navigator they'd taken on board at Morrison Army Airfield in Palm Beach. They'd refueled, pulled out and trimmed the plane, switched to autopilot, and in the blaze of early day, winged steadily south/southeast at cruising speed to punch the upper Bermuda Triangle. 1,329 miles ahead lay Puerto Rico, where they'd touch down and fuel again.

The two spoke desultorily of their flight out of Fort Wayne, then fell silent under the roar of the engines. At length Jack spoke.

71

"To think we got a whole ocean to haul this bird across. Numbs your mind."

Ned folded the map spread over his knee, and pushed it into the pocket by his leg rest.

"Don't forget South America and Africa. The ocean's just the fun part."

"Well, pray for smooth flying, Ned old buddy. We got a passenger aboard we don't want to scratch or rough up, not if we figure to keep our hides. And I'm not talking about Hap the Gazer back there asleep with his charts. Hey, that mechanic at Morrison had the laugh on us."

"Guess so."

"We knew the name. Just never thought about spelling it."

"Heard it in the newsreels plenty of times," Ned agreed.

"Well, listen up, pal, 'cause I've got this business covered. Our respected rider, and I am not referring to Dome Boy out cold there at the desk, but that spanking new engine in the cargo bay. Which as we know, happens to be for a '46 just like ours. Now say we get unlucky and lose an engine. We just pull down someplace off the track, round us up a mob of buck-naked natives to uncrate that baby, and make a little switcheroo in the dark of night."

"Gotta hand it to you, Jack, you're thinking."

"Only what if we lose another engine? Well, no one lives in a perfect world. Bet that grease monkey back in Florida is still laughing. Whatcha' think?"

"I'm the one with egg on my face." Ned cracked a wry grin. "Trying to sound out the name where the engine's headed. Mi-kee-nah-ha—."

"Guy laughed his head off. *Fellers, that thar engine's a-headed fer ol' Michinaw, whut Vinegar Joe Stilwell done driv' out them Japs. Don't you fly-boys know nothin'?* I felt like a pinhead. Not that I was ready to tell him so."

"Well, we file the flight plans, so we get the bead on where we're headed." Ned paused as his mind raced to the name on the crate, Myitkyina. The inner flash that told him they'd be flagged along to India. And from there, unless he was wide of the mark, on to the Hump.

He turned to Jack. "Plenty of men leave with sealed orders. From what I hear, they're on a troop plane, open their papers out in the middle of nowhere, and don't know what hit them. Mayer, as you live and breathe, we're roaring down the Fireball to the end of the line. Stake your last dollar on it, we're going to see China."

* * *

The long strung-out supply line to the Hump formed one of the most amazing roads in the history of the world, and it was in the sky. A circuitous course across the globe, it integrated former air routes and forged new ones, roughly 10,000 miles from the continental U.S. into northern India. Few pathways in the history of the globe, short of China's old Silk Road, ever proved more challenging to thread, yet so revolutionized the movement of goods and men. The western terminus had originated as early air travel sought to bring propellor aircraft across ocean distances with adequate provision to refuel. With flights to Asia across the war torn Pacific being out of the question, a shorter so-called Crescent route had been developed to span the North Atlantic from Dover on the eastern seaboard, through Newfoundland, the Azores, then Africa. There was, as well, a less-used Snowball route to the UK, farther to the northward.

The southerly Fireball, portions of which grew around early-flying Pan American Clipper and mail routes, had been reconfig-ured in 1943, at the instigation of General Harold George of the old Ferrying Command. With recourse severed to the Middle East by way of Europe, and the North Atlantic crawling with German boats, a trail had to be blazed skyward via the South Atlantic to meet war needs in CBI. Ground crews and fueling stations at strategic points made it possible for pilots to ferry airplanes, servicemen, supplies, aircraft parts, and vehicles from the continental U.S. to Dinjan, India, in a matter of days rather than months—a Pony Express on wings—if enough pilots could keep at dead heat. Ned's hunch about how far they were headed was based on clues provided by their cargo, but typically, men left south Florida without knowing their final stop or what assignment awaited. Hump pilot Don Downie writes of his own group of ATC pilots' being transported over the Atlantic in a C-54; not till they were in the air and opened up their orders, did they learn they were going to India. [17]

As the sun dropped into the western rim, two weary flyers touched down at Borinquen Field, in Aguadilla, Puerto Rico, ready for a stretch, a meal, and a bunk. But news that awaited them that evening of April 29, drove sleep from their eyes and hunger from their craws. Germany had just surrendered to the Allies. World War II was over in Europe. Which put the score now at half-the-world down, half-the-world to go. At least the ball was moving.

Next morning before takeoff, Ned headed out to the revetment, circled their plane on foot, and checked it up and down, a habit he never neglected. He kicked the chesty bulk of the tires —ground crews could always miss something. Then they were back in the air on instruments, flying a steady route south along the Fireball skyway, another 645 miles till they put down at Atkinson Field in Georgetown, British Guiana, to top the tanks. Generously, he hoped.

Here they officially attained the northeast coast of continental South America. Under their hands hummed a costly new airplane, its vital cargo, and a cat-napping, utterly superfluous navigator. Being little inclined to amenities, Uncle Sam left pilots to spell each other off, hold their course by instruments between beacons, and improvise weather reconnaissance with a searching eye, radio static, and zero reportage. No smiling crew chief stepped to the cockpit door to proffer a tray of pastry or cup of hot coffee. Alone between the far-flung reaches of blue heaven and cobalt seas, they unstoppered their canteens, traded seats, unfolded their maps, figured out where they were, and charted their course.

On May 1, they flew another 727 miles into Belem township, lying in the steamy clime of the Equator along a wide Amazon estuary into the South Atlantic. Following a short night's sleep, they shaved, rebuttoned their rumpled khakis, laced their brogans, and flew out at daybreak—1142 miles that day, to touch down at the airfield at Natal, a prong on the edge of South America's Horn that on a map calls to mind a figurehead on the prow of a ship. With a few winks under their belts, they took off again just before dusk in order to fly overnight their initial leg across the vast spread of the South Atlantic.

In the dark of these hours, their navigator, Lieutenant Beckman, blinked to life. He unfurled his maps and unpacked his compasses. Climbing atop the desk in his shoes, he thrust head and shoulders into the dome at the crest of the plane and commenced to "shoot the stars." His heaven-gazing made mainly for interesting talk, as they'd flown by radio compass since leaving Brazil, steadily on course until the Natal ADF wavered and died, after which came a quiet, intense interval till they homed on the signal from the Ascensions. Their passenger, they learned, served a stint in Europe before his assignment to ride the planes to Karachi, one of a group of navigators sent along the Fireball to chart the way for pilots who lost their signals over the open water, a hazard that created extreme apprehension for many of them.

It was by instruments and their radio compass, located on the top of the plane like a spear just forward of the dome, and a strong signal, that they hewed to their course. Few sensations on earth equaled being a microscopic mote alone somewhere in the night sky above a blackened, bottomless sea. No lifeboat, nose-ahead into the hard fists of the trade winds of the Southern Hemisphere, the sheer isolation of the flight made pilots the more appreciative of their ADF. This navigational marvel —the Automatic Direction Finder—picked up the beacons on which they homed inside the cockpit. Its critical dial with a slim needle, dubbed the Bird Dog—always a pilot's best friend—told them if they strayed to the right or left of their signal, or kept dead-on. Considering the vastness of the South Atlantic, the tireless blast of headwinds and absence of all earthbound markers, the need to hit the microscopic dimensions of a tiny island from millions of cubic miles of sky, the instrument seemed a wonder and a miracle, and probably was.

At latitudes below the equator, air currents predominate from the east or east/southeast, so Ned and Jack were working against tremendous winds. Another man headed for the Hump, in fact flying the same route during the identical April and May, 1945, though with several years more experience, weather pilot Otha Spencer, describes the end of Natal's runway being strewn with "radio gear, gun boxes, and other items thrown out by crews to lighten their planes before taking off," so great was their dread of ditching in the South Atlantic. While Ned terms his own ocean flight uneventful, or at least not dire, Spencer, despite his relative maturity, terms the run "pure hell," and confesses that he and other flyers nearly jammed the airwaves trying to stay in radio contact. "They just wanted to know that they were not alone over the Atlantic. Few knew where they were going. The miracle is that it all worked out."18

Pilots had to reach the succeeding terminus at Ascension before daylight to time their arrival when the seabirds left the runway to fly out and feed in the breakers. Sooty Terns or "Wideawakes," the fowl were called, and after the rising of the sun they clustered by such thousands on the tarmac that pilots couldn't set down their planes—a tough enough task in the half-light as they barreled down fast and steep between two volcanic peaks, fighting the wind shears of the sea and trying not to embed their heavy aircraft in the lava.

"A pretty desolate place," Ned assessed Wideawake Airfield, a crucial air and naval link, if whimsically named. Its modest seat at Georgetown presided over a volcanic island of craters and ash out in the equatorial seas of the South Atlantic Ocean. Ascension was about equidistant between the coasts of South America and Africa. Just to locate this pinprick was so difficult that pilots recited a morbid jingle, "If I don't hit Ascension / My wife will get a pension."

Jack and Ned wound up with an extra day on their hands when their plane's topnotch new left engine turned traitor and failed to check out. They'd flown 1310 miles of ocean, reasonably in the Commando's range, in pitch darkness. The island's total sequestration was a sensation in itself. Even the sister island of Saint Helena, where Napoleon Bonaparte ended his days in exile back in 1815, drifted another 800 miles to the southeast.

With new spark plugs firing robustly in the engine, they winged out again into the earliest sliver of sun, ever considerate of the terns' feeding habits, then high over the sea till at late day, they put down on Africa's Gold Coast in British Ghana. They stepped out to a hot wind blowing at Accra airfield, where having barely set foot on the Dark Continent, they napped and took to the air again, still dependent on maps and instruments, on signals by radio compass, and each other's know-how, as they doubled around to thread a westerly flight 800 miles into Robertsfield, near Monrovia, Liberia. The ADF was transmitting well, and from time to time they spoke to stations along the way, English being, as always, the language of aviation.

Weary hours later they attained Liberia. Here they cut away from the coast to start their overland flight above the continent's heart, steady to the north above French West Africa, where in a matter of hours, the plane would fling its winged shadow over the vast Sahara, still wheeling along the endless Fireball causeway in the sky, till deep in Morocco they put down at Marrakech.

* * *

The month of April 1945, wrote some significant world history. On the 12th President Roosevelt died, and his successor, Harry S. Truman was sworn in. The 26th ushered in the opening of the United Nations Conference in San Francisco, where fifty nations would sign its Charter. The 28th saw the capture of the Italian fascist dictator Benito Mussolini, who was hanged on the spot by his countrymen. April 29 heralded Germany's surrender. The next day, Hitler committed suicide in his bunker in Berlin, as Soviet troops swarmed in to occupy the city.

Drilling into the Fireball skyway, Ned Thomas and Jack Mayer were approaching a side of the globe where no one could view events with any such measure of conclusiveness. While they flew steadily eastward, U.S. troops overseas began the invasion of Okinawa, another island war against a foe simultaneously attacking Chiang Kai-shek's various strongholds in China's heartland. Japan's rationale for the stepped-up aggression on the mainland was said partly to result as a response to the introduction in China, in 1944, of the world's largest bomber, the four-engine Boeing B-29, which in due course was to fly under the banner Enola Gay.

The B-29 Superfortress by any name was a terrifying plane. It appalled the Japanese, and helped ignite the Ichigo offensive which inflicted new levels of suffering and brutality on the Chinese, who already were abused past endurance. The eventual mission over Hiroshima and Nagasaki was to bespeak all the cruel trade-offs that confront human beings in time of war.

That being said, in CBI the Superfortress proved a high-priced and death-dealing disappointment, another fiasco in a region that routinely defeated the most advanced aircraft. Brought into the theater under Operation Matterhorn, at Stilwell's insistence the plane had been based in India, another forlorn hope not to provoke the enemy. Stilwell also stipulated that the B-29s be re-staged from inside China, a convoluted plan that forced added fuel to be flown over the Himalayas to relaunch them from their Chinese bases. How many pilots flying that "Skyway to Hell" must have called the bomber accursed. No amount of tonnage quelled its voracious appetite. The '29 siphoned off three-eighths of total fuel supplies in a territory where every nut and bolt, every wheel, propeller, plug and drop of oil, had to be flown over the highest range on earth. A quasi-arrangement quickly evolved whereby, in addition to supplies gobbled elsewhere, the bombers flew some of their own fuel.

Because the B-29 needed runways a mile long simply to get off the ground, Chiang Kai-shek set out to provide new airfields. Pavement had to be hard-topped because the older gravel, dirt, or metal-mesh arrangements that made do for other aircraft couldn't handle them. In the greatest call to Chinese labor since the construction of the Great Wall, peasants by hundreds of thousands swarmed out leveling, excavating, building from the first chink of light till darkness. They used bare hands and little more than stone-age tools. In baskets slung over their backs, they carried off the mud of the rice paddies, and brought back from river beds the rocks they broke apart with mallets to lay for the foundation. At nightfall they dropped where they stood under shooks of straw they pulled over their heads for shelter from rain, wind, and cold.

The Generalissimo probably felt all the effort had been worthwhile the minute he saw the Superfortress fleet wing in. The plane was heart-stopping. In length 99ft, having a wingspan of 141ft, streamlined fore and aft to eliminate drag, the sleek bomb-shaped fuselage had flush-riveted skin, and even the gun turret was remote-controlled. Cruising at 358 mph, it carried a payload of 20,000 pounds armament, plus a dozen machine guns, 11,500 rounds of ammo, and a 20mm cannon with another 100 rounds. Its long legs gave it an astonishing range, 5600 miles, and it pushed against an impressive ceiling of 31,850ft, having one of the first pressurized cabins in aviation. What wasn't to love?

Plenty, as it turned out. The B-29 imposed new supply and

logistic hardships in a part of the world that had nothing to spare. In the blistering temperatures of India, the four engines had an appalling proclivity to overheat, and when this happened they crashed. The Superfortress also overheated in cooler China. Some hardly cleared runways before exploding, others blew up trying to land, or had to turn back en route to their targets as mechanical failures spun the dials. The aircraft required a crew of eleven, all of whom had to be meticulously trained, though for a prolonged time, the B-29 was not produced in sufficient numbers to provide trainers.

Modifications were jury-built on site to cool the engines more efficiently. Trained pilots under Operation Matterhorn had difficulty simply navigating with sufficient precision to hit their targets. Barred from flying in formation, the planes were more vulnerable to enemy counterfire. With Washington superceding commanders on scene, exigencies arose which could not be dealt with in a timely way. But maintenance and supply nightmares really sank the B-29 in China. Chiang Kai-shek, standing proud and gratified to see his American friends sending out the great bombers in his cause—for the B-29 was used only in the Far East—found his vainglory sadly muted by the fact that as long as the Superfortress ate fuel, his armies had to make do with less.

After some twenty inconclusive missions in China, the B-29 had accomplished nothing to break Japan's will to fight. Indeed the plane seemed to inspire a gritty redoubling of hostile effort. General Wedemeyer ended the business and had the Joint Chiefs withdraw and write off the '29 for CBI—a $3 billion blunder, and who ever heard of such a sum? But the lessons were being weighed hard as the planes relocated to Tianin Island near Saipan, where their role in the war would be assessed with valuable CBI experience to draw on.

The departure of the B-29 from the China-India-Burma Theater also made clear that the hefty bull of a plane Ned flew, the C-46, would prove, as he said, the "heavy hitter."

* * *

The 1807 miles from Monrovia to Marrakech kicked the legs of the C-46 out to the max. The plane's range was engineered at perhaps 1600 miles, though figures come in as low as 1200, perhaps to allow for outsize cargos and the exceptionally harsh conditions of the Himalayas. For the Fireball route, Ned and Jack were flying with a minimal load, and a heavy engine to transport wasn't much to ask of the Curtiss Calamity, as the aircraft also was called. Over Africa they no longer had the ocean's headwinds to contend with, and all in all, conditions were favorable as they passed above the timeless deep-green swaths of jungle that gradually faded into the paler coloration of the savannas, then dried in the vast shifting whorl of the desert.

Four years Ned's senior, weather pilot Otha Spencer would call it beyond belief that "young boys" like himself were flying routes for which airline pilots took years to qualify. "The airplane," he writes after his own highly anxious run along the Fireball, "was truly a magic carpet, a miracle that is every pilot's dream—flying halfway around the world!" [19]

Ned and Jack set down in Marrakech, only to have the other new engine play them false, though it held for a landing. Forced to quarter an extra night for more plugs and repairs, on this their second delay, they were not amused when another plane out of Palm Beach overtook them, though it departed a day behind. The temperament of the Commando was showing itself, not for the first time. Back in November of 1943, when the first C-46s reached CBI, 31 immediately exploded in flight. Others gave up the ghost in assorted ways, or had to be sent back to the States for retooling, with a punch list inclusive of some 721 corrections, a figure pilots shook their heads over and cursed. The plane's reputation for misconduct also came as a result of its having been hurried along production and inadequately tested on stateside mail routes, or calmly over airfields, hardly preparation for what they'd be subjected to in the Hump. Blame for the inadequate engineering would be laid by historian Barbara Tuchman on the Generalissimo's "Dragon Lady," Madame Chiang Kai-shek incessantly pestering Roosevelt for new planes before their reliability could be determined.

Hump pilots flying their grueling runs took the C-46 through the inundating monsoons of India, in cockpits that leaked so badly that they often flew in soaked clothing as water sluiced down their collars, a situation that earned the rugged plane another epithet, the "Plumber's Nightmare." There were, too, the disastrous ground-roll landings when big "Dumbo" seemingly wanted nothing more than a wild abandoned wallow in the mud—and a ground roll under so sizable a beast always threatened to be on the lethal side. Assorted glitches ran up and down the arteries of the hydraulics, while spark plugs, indeed all manner of sparks, were a byword in the aircraft, with leaky oil drums and stray discharges from the electrical system often blamed for explosions. After the war an unvented wing structure was discovered to be another culprit in electrical build-up, and wing vent modifications eventually were incorporated. The list goes on, but whatever the problems, pilots paid with their lives. No doubt about it, the plane's appellation as the Commando was hard-earned, yet proven in the end, as a sturdy giant that could show its mettle under the harshest tests put to men and aircraft.

At Marrakech, two sleep-deprived pilots quartered in and slept hard. The blowing sand, the camels, the water sellers with their heavy bags slung on their backs, were a novelty for two fellows from North and South Carolina respectively, and they certainly caught Ned's eyes, but did not keep them open for long. Hard to think this pilot of twenty-one, stretched out on a cot in a hot bunk room, would fly into the city fifteen years later, the sand and the water sellers much as they'd been from the time of the Patriarchs, and be given a suite as big as a house at the luxurious Hotel Mamoussa, where Winston Churchill stayed while he painted the Atlas Mountains.

Now it was Cairo ahead, 1425 miles to the eastward above the vast, blowing Sahara. Flying low in clear weather—for enemy fire no longer menaced pilots across North Africa's Maghreb—Ned looked down and saw tanks and munitions strewn along the sands where they'd been abandoned in retreat by the Desert Fox, German Field Marshal Rommel, who'd paid with his life for his part in the conspiracy to kill Hitler.

At Cairo came another two-day stopover for maintenance. Ready for a bit of adventure, the men headed out like typical tourists to Giza, to see the pyramids and take a camel ride. Ned, who could gun an airplane through a chain of snap rolls with guts of steel, found the back-and-forth lurching not to his taste. "Never ride a camel," he maintains solemnly, "unless you practice a lot. It'll rub your skin off." An old snapshot shows the men in their khakis, Ned looking into the camera alert and good-natured, Jack with his hat rakishly angled to the side, the evident ease of the two pilots in contrast to the terse stance of their navigator, who stares down at his beast nervously, a riding crop lifted in one hand.

Next destination, Abadan on the Persian Gulf, clocked in another thousand miles. Taxiing to a stop at the Iranian port and oil depot, Ned opened his cockpit window and got hit with the suffocating blast of a furnace. The Fireball route seemed better named with every mile.

"I felt that blaze and thought my engine was on fire." They'd crossed the Sahara south to north, west to east, and never experienced heat like Abadan. It was hard to rest in the stifling tent that quartered pilots, but they hewed to routine—sleep, eat, refuel, fly. 1141 miles lay between them and their next terminus, Karachi—another place to set a plane down along the unfurling Fireball racecourse, though by no means the end of the road. Not by a long shot.

* * *

As the Hump airlift winged forward, mounting achievements moved another statistic up the charts. Fatalities. In the first 90 days of the present year, January to March, 1945, 70 planes were reported downed, taking the lives of 134 pilots and crew, a figure later reassessed as incomplete. On the China side during the same three-month span, Chennault notes the successful targeting of a million tons of Japanese supply and the downing of dozens of enemy planes, all without loss of a single flyer in his command. Why did his men have so markedly greater a chance of survival than pilots flying the Hump? Bluntly put, Chennault's fighters weren't taking outdated aircraft through the stratospheric conditions of the Roof of the World.

Unbowed by their recent defeat in Burma, Japan mounted a series of barbarous new offensives. Aiming to hobble the Hump and capture China's largest terminus at Kunming, a lifeline both for Chiang Kai-shek and Claire Chennault, Japanese troops stormed and took Suichuan, Kanchow, and on March 25, Laohokow, all in eight weeks' time. Desperate for supplies and planes, Chennault looked more than ever to the Airlift, but as in days past, found himself forced to stand in line, this time while pilots flew Chinese troops east to stanch the bloodletting at Chihkiang.

General William H. Tunner was a few months into the command he privately viewed as the graveyard of his career. His consternation had cause. Caught between Washington, DC, and the loud wails of the Nationalists who depended hour to hour on the men who flew and died crossing the murderous range, would he wind up like other generals in CBI who lost their health and sanity, if not a star or two? Of this "unprecedented operation at the ends of the earth," he writes, "For every thousand tons flown into China, three Americans gave their lives. . . . Not only was the accident rate alarming, but most were washouts—total losses, with planes either flying into mountain peaks or going down in the jungle. In many of the cases in which there was reason to believe that some or all crew members had been able to parachute from their planes, the men were never seen again. The combination of a high accident rate with the hopelessness of bailing out was not conducive to high morale. It was safer to take a bomber deep into Germany than to fly a transport plane over the Rockpile."[20]

It is fortunate that Tunner quickly saw his role as a constructive one, not just to drill holes in the blue sky but to save life and make the posts more livable for his men. He certainly knew that when bases are orderly and run in soldierly fashion, productivity increases. So how did he find Americans subsisting in the shadow of the Rockpile? Sending his aide Hammie Heard ahead into the Assam Valley, he got a report:

"Sir, I have never seen or heard of people living—existing— as they do here and out on the operating bases. This is grim. Nothing counts but the tonnage that goes over the Hump. Morale is at the shoelaces."[21] Does Heard overstate? Not according to a wire General George E. Stratemeyer, commander of Allied Air Forces in India and Burma, sent Major General Harold George (a copy went to Tunner). "I am thoroughly sympathetic with your desire to achieve the finest safety record possible on your Hump operation. However, I feel that you have given flying safety first priority over tonnage production. Particularly restrictive has been your insistence that the number of accidents be reduced even at the expense of essential tonnage. I recommend that you remove the stress on the number of accidents."[22]

Some sympathy, a wonder Stratemeyer signed his name. He and General George even argued whether to reconsider the Hump's combat status across the board, though in fact pilots already flew the theater under combat ratings. Yet purely to have the astronomical fatality rates assessed more leniently, the two kicked around the idea of calibrating deaths against time flown rather than raw numbers, inasmuch as a single Hump run involved long hours of flying.

When General Tunner reached India and toured the bases for himself, he was not edified. He found men living in mud and rain, plagued by malaria and other jungle miseries including a pernicious fungus called the Dhobi Itch. They went unshaven due to shortages of razor blades and soap. Post exchanges routinely ran low on the most basic personal supplies. Toothbrushes and toilet paper got rationed or raffled. Shirts were on backorder. At Thanksgiving when American troops around the world were promised a turkey dinner—news of the feast was blazoned across the Armed Forces networks—his men got wieners and kraut. "Where was all the rest of it?" Tunner roared. "Ha. This was the Hump. We were at the end of the line." 23

He found his men lacking access to the simplest outlets to pass the time. Ice, candy bars, magazines and newspapers, were scarce to nil. Pilots grew gaunt on tasteless indigestible rations. Diarrhea and dysentery were rampant. Tunner encountered men not simply burned out, but who really were sick. Small wonder they grew indifferent to shabby unlaundered uniforms. "Officers and gentlemen" swatted vicious insects, scuffed around stifling airfields in shorts and chukka boots, and at dinner wore garrison hats defiantly askew, with unbuttoned shirts and unpressed trousers. They flew with stubbles and mustaches, overdue for haircuts, fed on Spam, hash, grits, anemic margarine, green eggs, and potatoes reconstituted in brackish water—water which would have made any U.S. health department nauseated. Tunner inspected and saw dead bodies floating above the source of supply. "Water used for sterilizing dishes in the various mess halls is, in fact, a form of soup with a basic foreign element of well-cooked feces and red mud." He said he personally wouldn't wash his feet in the stuff.

He encountered an arresting picture of men who'd lost all respect for military discipline. Enlisted ground crews didn't bother to salute officers and commanders. Tunner met up with men who did not bestir to stand at his approach. As for pilots, little they cared how they appeared while fatalistically they fought to survive under the insane rule, no weather exists over the Hump, no matter what they went through. Most hoped just to stay alive, log in their hours and get home.

Saying his flyers looked like "a bunch of damned taxi drivers," the general laid down the law. "I was fortunate in that I had the opportunity to put my finger on this problem of morale. And I knew what to do about it." [24] What he did was re-institute daily inspections for the enlisted men (weekly for officers) and formal dress parades on bases. He cracked down on shaving and uniform correctitude, ordered regular hair-cuts, and cleaned-up living conditions generally. He attacked the pestilence of mosquitoes by cutting loose some retrofitted B-25s with DDT sprayers. Skeeter Beaters, he called them, and they drove down the malaria figures.

"Now I had the opportunity to take over a chaotic situation and make something out of it. I never doubted, not ever, that I would succeed." [25]

Nothing loath to blow his horn, Tunner couldn't improve the food much, but he made more available, kept messes open for all-hour schedules, and mobilized snacks and drinks to the flight line in PX jeeps. Pilots were to be driven to their planes in jeep taxis. "Quiet" posters were installed around quarters so they could sleep after flying through the night. Men grumbled about a general who expected them to risk their lives every day, then march in formation like raw recruits through filth and mire, and salute while bugs chewed their faces. But Tunner didn't back down, and morale and discipline inched up, probably more than anything, due to his recognition that there was, in hard truth, hellish weather on the Hump.

"There was weather on the Hump, and we had to face the fact. I knew that our mission was to fly tonnage to China to prosecute the war as vigorously as possible and that therefore we had to fly under tough conditions. We did so; no airline today would permit flight under the conditions we worked under." 26

Crews suddenly were under orders to turn back if conditions got too dangerous. Though as in many matters, it didn't take long for operations officers to ignore the protocol in order to keep schedules going. Pilots, too, often shrugged and flew on with gritted teeth, knowing that even if the weather got better, it wasn't getting much better.

Still, Tunner continued his shakedown. Glowingly in his book *Over the Hump*, he writes, "Though the first parades may have been composed of slouching, self-conscious, and diffident men, you could see each Saturday, a little more stiffness in the backbone, a little prouder tilt to head and shoulders, a more purposeful swing in the arms. They were beginning to look like American soldiers, and that was just exactly what I wanted. I had been sent to this command to direct American soldiers, and while I was their commander, by God, they were going to live like Americans and be proud they were Americans." 27

On a practical note, he recognized that planes were being overloaded, often by as much as 3000 pounds—a problem all his orders did not effectively conquer—and that many were badly out of balance besides. Load masters were under orders to use extreme care in distributing the tremendous weights, which on occasion got heisted on aircraft by powerful elephants guided by their Indian mahouts. Of course, when cargo consisted of unruly milling foreign troops, all this became pretty academical. Ned himself would constantly recheck loads and tie-downs, and it usually was apparent to his own eye that safe limits had been leniently construed, to put it mildly.

Tunner found pilots flying under severe operational fatigue. His biggest reform may have been a rule raising flying hours for rotation from 650 to 750, or a year regardless of hours flown, and prohibiting in excess of 65 hours per month at the controls, the maximum he considered safe. The intent was to eliminate the incentive to fly exhausted or impaired to get home sooner, factors that also contributed to casualty rates. Over time, fewer bodies got shoveled off the ends of the runways, that was the good news. But a new broom sweeps clean, and CBI didn't beg the rule. Ned never saw a dress parade at any base, in China or in the Assam, and as for Tunner's minute inspections, if they happened at all, he must've been in the air. Long flying hours continued, too, they just weren't logged in. Ned and others flew schedules dictated at operations, kept going, and never blinked at the happy publicity of a general living a thousand miles away in Karachi.

On arriving in India, so as to gain a better understanding of what his pilots experienced, Tunner recklessly commandeered a C-46, a plane he'd never flown, and took off with his startled staff over the Hump to Kunming. It was a rare clear day, but the escapade could have got them all killed and nearly did, a lesson driven home as the general looked down on the runway at Chabua, and saw the large blackened areas where other planes had crashed and burst into flames. Acknowledging he'd done a highly foolhardy thing, even so, his sympathy grew for his men. Bluntly he warned his own commanders:

"In striving for high aircraft utilization, we will not sacrifice flying safely. One hour lost can be made up later. The loss of one load of passengers and crew can never be recovered."

* * *

At their stopover in Karachi (now in Pakistan) the engine in Ned's and Jack's cargo was offloaded to a plane bound for Burma. Their navigator said good-bye and headed back, happily no doubt, to Palm Beach. The pilots reported to the records clerk and were not surprised to learn their assignments to the Tenth Air Force at Dinjan, farther north in the Assam Valley. In the interim the two had a couple of days to spend at Karachi's landscaped officers' club and do some wandering through the ancient city.

They opened their eyes on a new universe in the color, beauty, and squalor of what was then an Indian protectorate on the Arabian Sea.

British colonial architecture spread its faded bloom along wide streets and inviting parks in prosperous districts where dwelt well-dressed Europeans. Traipsing farther, Ned met with the wave of native humanity, men with heads wound in turbans, women in thin saris nursing naked infants and trailed by naked children. For the first time in his life he saw beggars. He gazed on the fabled snake charmers, he viewed the ornate pigeon huts and pleasure gardens along the water. The roads teemed with donkey carts, camels, bullocks, horse-drawn carriages, and hand-pulled rickshaws. Ablaze at unexpected junctures were the funeral pyres of the dead. He wandered into a bazaar and got fitted with chukka boots to ease his feet from the hard footwear of duty hours. He dug deep and bought Neva Rae a ring with a handsome sapphire solitaire, which proved later to be worthless runway glass, not an unknown experience for a trusting American serviceman in exotic parts.

When they got in the air again, it was to traverse the width of the Indian subcontinent, 1342 miles to Calcutta, a place which even after Karachi, appalled and sickened Ned, for he never had witnessed poverty and human suffering on the scale he saw in this teeming place of poorest people and sacred cows, with its rigid castes, famished children, wasting beggars, and destitution. For a short stopover, the place registered a lasting impression, but soon they were winging another 400 miles north over territory that would become present-day Bangladesh, and beyond this, far into the Assam valley of India, where they set their plane down at the airfield at Mohanbari, a green land of tea plantations and cultivated hillsides, lying below the forward slope of the Himalayas that crowns upper India, shoulder to shoulder with neighboring Bhutan.

Leaving Dumbo, Ned and Jack caught a military plane and

"deadheaded" northeast to Dinjan. Crews welcomed newcomers, and from the open cockpit door, they soon heard the guiding voice of the Dinjan control tower, a ragtag-appearing structure standing amid bobtails of tropical grasses and a long paved runway that sprawled near the flood plain of the Brahmaputra River, a major line of demarcation before the rise of the Himalayan foothills. The tower, framed up of timber and fronted with a bare race of metal stairs with a landing, was roofed in thatch like most buildings on the compound. With its loudspeaker emerging on top as from an unkempt wig of fodder, it made a raucous reception for pilots descending into the hubbub of a working air base, present home to two U.S. Troop Carrier Squadrons in the Tenth, their support and maintenance personnel, and native help. The roar of the planes flying in and out made a din heard round the clock.

Dinjan, one of a cluster of airfields dotting India's Assam Valley, and the first to accommodate larger aircraft, took its name from a tea plantation located at the terminus of a narrow-gauge coal fired railroad, which still chuffed in with supplies. Across the Hump, 560 miles away, lay Kunming, China, and about 40 miles southeast, Ledo, where that recent January, Stilwell's perilous sniper-plagued road first embarked on its torturous trek into China.

Pilots new to the Assam discovered a steamy-hot, snake-infested, primitive, insect-ridden place, a flyspeck in the crease of the map, and at the time of Ned's arrival, soaked by seasonal monsoons that originated from the Indian Ocean and south Asia. These weather patterns blew in from the southwest to blanket the valley in boiling, water-laden clouds and mind-boggling torrents of rain. No hangars sheltered aircraft at Dinjan, and maintenance staff serviced them as with the secrecy of gremlins in dark of night, for the metal skins of the planes were too hot to touch in the sun. Ned could not afterward remember ever seeing a plane under the hands of work crews in the heat of day, for all General Tunner's pushing of production-line servicing throughout the

Hump. During raids, with proper revetments lacking, the only cover for planes involved taxiing them off the field and dispersing them into the bush, assuming they had warning—which seldom they did—and it was hard to camouflage large aircraft anyway.

Ned's status changed from transit to his wartime assignment with the 1st Troop Carrier Squadron of the 443rd Group, in the Tenth Air Force. Three squadrons operated under the 443rd—the 1st, 317th and 322nd. Dinjan was home to the first two, and each occupied its own section of quarters. He saluted his commanding officer, Major John V. Langford, a former flight officer who headed up the First Squadron, a wiry energetic man with a clipped mustache, whom Ned later would accompany into China. No Tunner-style debriefings followed the salute.

Like all newcomers, he got issued a Colt .45 to be carried under the left arm in a shoulder holster with a strap to anchor it to his belt—"they didn't want a sidearm out there flapping in your face if you had to bail out"—a leather flying jacket, and a trench knife, handy for opening rations or beer cans, and which would cut him loose in the event he had to bail out and got himself tangled in the upper limbs of the Burmese jungle. Accounts of untenable horror were told of pilots who neglected to pack this knife and got eaten down to their skeletal remains by ravenous ants, as they swung upside-down from parachutes caught in the treetops.

Ned came to regard even the pre-flight check-over of parachutes as a joke.

"The pilot's seat was built with a space beneath where you could sit on the thing," he describes, "but usually I stashed mine back in cargo and left the cushion in the chair. If you run into a rock pile, you won't need it anyway." Since General Tunner himself acknowledged rescue prospects for downed pilots at zero, who could blame him?

Men at Dinjan were housed in low stucco dormitories with open windows, straw roofs, and cement floors impossible to keep free of mud. They slept on canvas cots under mosquito nets. "Just long huts with a thatched roof," Ned remembers. "We called them bashas, an Indian word. They probably had been used to sort tea or to house the pickers, and were holdovers from the days of the plantations. The weather was so wet that you carried everything you owned on you, even your hat, because if you put it away and pulled it out again, it was green. Mold grew on clothing, belts, everything."

Olive drab sheets and laundry were tended by a houseboy called a walla. Upon rising, men upended their footwear to rid them of any scorpions that perchance had nested inside during the night. They saw the creatures running across the bashas, big as crayfish, more dreaded than crocodiles, and carrying the sting of a viper.   Compared to the living quarters, the primitively thatched base control tower was a wonder of modernity. Clothing was strung along rods, as even underwear grew moldy folded away overnight. Tropical snakes infested the region, so beds and bags were always checked. For the fifteen or so sharing a basha, there was a cold faucet and a cold-water pipe at the far end for a shower. Thanks to Tunner's reforms, an Indian barber came regularly to trim their hair using an old-time razor, after which he'd snap their heads from side to side to get rid of any "cricks" in their necks, a finishing touch favored by native clients.

In writing a piece for the Hump Pilots Association Newsletter (Autumn, 1986), veteran Hugh Crumpler describes CBI as "the only war theater where a soldier could be felled by a Jap sniper, bitten by a poisonous snake, trampled by a rogue elephant, decapitated by a headhunter, and devoured by a man-eating tiger. And that's just on the way back from the latrine."[28]

It didn't take a long memory for new arrivals to absorb the short and violent lore of the Hump Airlift. The Japanese posed the principal menace—a reduced one in the air after the Ichigo offensive burned off its fury, but men had greater cause by far to dread them on the ground should they go down in China or over the hills and jungles of Burma. They were literally everywhere. Tokyo Rose had just broadcast her message to Hump pilots, that any falling into their hands would be executed mercilessly on the spot. The Communists under Mao Tse-tung, by contrast, only promised a long, mean captivity while they awaited ransom. There were besides, various aboriginal tribes, in particular India's Naga headhunters, though perhaps no one knew for sure if they carried out their macabre artistry on pilots who went down and disappeared in tribal areas. Some fierce nomadic groups, also hostile to Americans, inhabited northwestern China, and one cannot neglect the cannibal tribes of Burma's well-nigh impenetrable jungles.

As much as anything else, men still spoke of what they called the Wildest Night, the "Night that Closed the Hump," when the enemy had been the weather alone, the worst they'd ever seen. Some of the pilots presently at Dinjan had lived through it, or remembered men who hadn't. Of this terrible day and night, January 6-7, 1945, General Tunner writes,

"The worst storm we ever had on the Hump occurred earlier than the usual spring thunderstorms, in January, but its characteristics were the same. It brought violent gusts and updrafts along with severe icing, sleet, and hail. A one hundred mile-per-hour wind, howling across our east-west routes directly from the south, blew planes far to the north among the high Himalayas. Conditions were the same from fifteen thousand feet to thirty-eight thousand feet—we couldn't get under, and we couldn't get over." [29]

Hump pilot Don Downie, with the ATC out of Chabua, was among those who flew that night:

"The weather was a very dense occluded front that moved across Africa and India, picking up energy all the way. As the front approached the Bay of Bengal, it picked up the needed moisture. There are no weather forecasting stations in that area, and the front moved in so fast that aircraft had already started out before the worst of the weather reports came in. When it hit the Hump, it really went to town. It had everything from hail stones the size of golf balls to winds that shifted to all points of the compass and went up to 150mph. Vertical currents were between 200 and 300mph. The cloud tops went up to 40,000ft and higher." [30]

Commands flying in the theater—the Air Transport Command (ATC) "who will fly when birds will not," the China National Aviation Corporation (CNAC), the China Air Task Force (CATF), the Tenth Air Force—all suffered appalling losses. Next day, it would be said, pilots could find their way across the mountains by following the smoke of the burning wrecks. The Rockpile had dearly earned its other moniker, the Aluminum Trail.

How many men died in the Hump? Records were kept piecemeal, sometimes not at all, and never tallied. It took the determination of one pilot's widow, Chick Marrs Quinn, to devote nearly ten years of her life, and read in excess of 60,000 government documents, to come up with figures which even by her reckoning are incomplete, as she states in the epilogue to her book, *The Aluminum Trail*, published in 1989. "One thing the reader must understand, there were very few reports filed for aircraft lost before June of 1943, and the reports were not filed for ALL aircraft lost. There can never be a complete book of all aircraft and all crews lost in the CBI. This information does not exist." Despite the limitations she worked against, Quinn documents the deaths of some 3500 flyers, inclusive of 1000 bailouts and missing, during the 3 years and 8 months from December 1941, to August 1945, among whom she lists the death in February, 1945, of her husband, Lieutenant Loyal Stuart Marrs, while flying aviation fuel in a C-109, the deadly converted B-24 Liberator tanker.

Among the latest deaths was a name that could not have struck closer to home for Ned. His best friend and roommate, the fraternity brother with whom he'd set out to share a dream to fly, 2nd Lieutenant Deming Ward, had gone down on May 8, also flying the notorious C-109. By then Ned had been winging along the Fireball into India.

How close their paths came to crossing. "Some went home." Ned searches his words slowly, then shakes his head. "If I hadn't excelled, hadn't stayed and instructed, maybe I'd have got there sooner, too. Maybe got out sooner. Or maybe been killed trying. You just never knew."

"This plane departed Dergaon at 0125Z, enroute to Kwanghan, China," comes the cryptic official report of Deming's fate, later copied into Quinn's Aluminum Trail. "They reported a position over Paoshan, at 0250Z, at an altitude of 17,500 feet, on Easy course. No other contact was made. On 18 May, 1945, Air Jungle Rescue notified Dergaon, that a native ground party had found the wreck. Three bodies were found with identification tags. (Pilot 2nd Lieutenant Demming) Ward, (Navigator 2nd Lieutenant Lloyd) Bjerk, and (radio operator Sergeant William) Ferguson. The plane was seen to burn and explode in mid-air on the 8th of May '45, and most of the wreckage had fallen into the river. It seems probable that the remaining personnel fell into the river, with the majority of the wreckage. It is believed there were other passengers aboard. We await a follow-up report."[31]

The account is lengthy for the Hump. Typically when a crash pitched into the range, no trace was found of the lost, no witness stood by to report or remember the final moments of another crew who'd given to their last measure for the Airlift.

Ned and his
mother, Anna
Rossie Pridmore
Thomas,
c. 1927.

Ned's first experience of flight: The Goodyear Blimp, Washington, DC.
Kiplinger Collection, Historical Society of Washington, DC,
Edwin Wisherd, Photographer.

Aviation Cadet Ned Thomas completes Primary Flight Training on the Stearman PT-17 at Douglas, Georgia.

Second Lieutenant Ned Thomas, Bergstrom Field, Austin, Texas.

Neva Rae Hudson Thomas, 1945. Photo stays in Ned's wallet through the war and through their marriage.

A C-46 Commando powers into the climb from India toward the storm-swept Hump. *Life* magazine, Sept. 11, 1944. Photo: William Vandivert. Getty Images

The Curtiss C-46 in flight.
USAF photo.

The Generalissimo,
Chiang Kai-shek.
Photo gifted to Ned
and other pilots in
China.

Hankow Airfield. Ned appears gaunt but composed; his battered
Commando the worse for wear, in this crackled old photograph.

CHAPTER 5

HOUSE OF SNOW

ON A DAYTIME RUN FROM DINJAN, INDIA, TO CHANYI, CHINA, the conclusion of two passes over the Rockpile, Ned checked through as a first pilot. The calm and cheer of his mustachioed Texan companion made good company while they flew by instruments, enwrapped in the twilight of the cockpit, the murk of the clouds, as wind shears and assorted caprices of water and ice buffeted the aircraft. They took a jostling, but nothing a self-respecting flyer was going to lose his head over. Henceforward Ned would work the round-the-clock schedule of the Hump, and captain his plane.

What of the helpful briefings ordered by General William Tunner for newcomers? A couple of check rides would be as much indoctrinating as Ned got, a glaring exhibition of unlikeness to the version Tunner left posterity. "We set up a routine procedure in which the new pilot coming into the theater would first take a short jungle-indoctrination course, then fly the Hump a few times as copilot, then undergo his final training for first pilot. For an adequate check on health and morale, the flight surgeon saw each pilot personally before takeoff." [32]

Exactly what Hump was the General talking about? What routine? What squadron? For sure not Dinjan, not the First, not with Major Langford. A jungle indoctrination? Forget it.

97

"Nothing was taken for granted," Tunner recites. "No matter if the incoming pilot had been a competent first pilot in some other division of the ATC, we made sure that he knew his job before he became first pilot on the Hump." With what softheartedness does he wax on the care lavished on his men: "The pilots diets had to be supervised. Eating gas-producing foods even hours before takeoff would result in debilitating agony at high altitudes."[33]

Ned never caught a glimpse of this solicitous flight surgeon either and wouldn't you know, his sole bout with illness came of one of these intestinal airlocks. The business cost him a few hours of agony on a cot in Kunming, after which he got up and flew his plane back.

He chuckles. "If the planes were ready, we flew them. It was a rotational system, 24 hours a day. In the Tenth we flew with a pilot and copilot. We didn't have a radio operator or navigator. A crew chief seldom flew with us unless someone wanted to pick up a few extra hours of flight time. Winds and rain often had us so socked in that I never saw the ground again till I landed in China. If we'd had to wait for a pretty day, we'd still be there. We were going through the monsoon season, when clear was the exception, not the rule."

Home-sweet-home for the Tenth's Troop Carrier Squadron, Dinjan, India, anchored a hard-driving operation where if a pilot was good, he exercised his self-reliance and acclimated fast. Unlike the ATC, another U.S. group, and most Chinese commands which flew with crews of four or five, in the Tenth pilots carried out work that in other outfits was handled by men on board. A crew of three, inclusive of navigator, was considered standard for the C-46, not that the protocol cut much ice in the U.S. Tenth, certainly not in the Hump, as Ned's experience underlines.

"I was my own pilot, navigator, radio operator and engineer, and I opened my own box of K-rations. I'd try to get the bacon and cheese," a twinkle lights his eye. "It had more taste and came with crackers. Or the spaghetti if they were out of bacon."

Not having a navigator kept pilots diligent, if they weren't already. The '40s had brought Long Range Navigation or LORAN, but in the Hump no one could use a system that involved vectoring two radio beacons from two stations over sectors where getting a signal from one station, much less a correct one, was a border-event miracle. Navigators liked working with LORAN when they could, and some such stations tracked aircraft in China from the ground by achieving their bearings in the sky, but this required a sufficient expanse of open territory.

General Tunner was not unmindful of the grueling routine his pilots lived. He just didn't aim for "frenzy." He did not want "flap." He wanted his planes in the air or the service bay twenty-four-seven, his crews either flying or catching their winks so they could fly again. He wanted a smooth operation under very rough conditions. Charts and numbers excited William Tunner, so long as they slanted up in efficiency, and down in casualties. He was aware, too, of the tendency of lieutenants to grouse, as well he should, having spent seven years of his own life at that modest rank, but fear not. "I had become known as a cold, hard driver, with the nickname 'Willie the Whip' whispered behind my back, and I didn't lose any sleep over it." 34

What kind of pilots did Tunner want? "Young pilots were the most prone to accident. Pilots of my age who had less than five hundred hours of flying time comprised the next class of men you wouldn't choose to fly with. Mature men with sufficient flying time but inexperienced on the Hump, came next. The safest pilot on the Hump would be a man in his thirties with two thousand hours or more pilot time, including at least two hundred hours in India or over the Rockpile." 35 Thirty-something? How many hours? At age 21, Ned had racked up a total of 208 hours and 20 minutes of cadet flying, plus operational work, with 90 or so added for spinning along the Fireball. The thing General Tunner failed to figure on was a strength that proved Ned's own, a cool head, an eye beyond his years for safety. He hadn't trained long but he'd trained well, and assimilated what he needed. Instrument flying was stressful. The major part of Hump work was on instruments with no visual reference to sky, ground, or the death-dealing surround of precipices and peaks. In turbulent mountain flying, a pilot literally might not know which end was up. Often he flew blind till he broke out of the soup (if he broke out at all) and saw the runway ahead to land his plane.

In work that called for self-possession under pressure, and where round-the-clock rotations made social time scarce, men were affected. Some drank hard. Most like Ned indulged a social beer. Some hesitated to form friendships with fellows who could go missing on the next roster, an attitude that fostered a some-what overworked notion that pilots were loners, silent cowboy types. For a few, there were poker games in the basha. For all, a swap of pleasantries in the mess. If nothing else, they could joke about the food. Dinjan's semi-circular base theater let them shed some of the stress of flying, and Ned remembers popular stars Erol Flynn, Charles Boyer, cowboy-crooners Gene Autry and Tex Ritter, Bing Crosby and his comedian partner Bob Hope, all of course, with the leading ladies. The need to focus and fly safely was something he viewed as precedent over personal ties, rather than a conscious intent to be reclusive. But at best, a pilot's life in wartime north India didn't foster camaraderie. Time was scarce. There was noplace to go. A restaurant in Dinjan plied a by-no-means bounteous trade, and now and then, a few would drive over in a jeep for chicken and vegetables.

"I gotta say it was a scrawny chicken. The place sat up there on stilts, pretty bare.

"Since we didn't have an officers' club at Dinjan, some of us got together and found an empty squadron room we called a club. We pitched in for beer and cokes, rigged up a sort of makeshift bar, and hung a parachute on the wall. Otherwise you could only sit on your cot. It wasn't much, but it was a place where we could go and chat, and just drop in. The men liked it."

The clubroom Ned instigated grew popular. He searched out odd chairs and tables, and empty crates to upend, innocuous adventures he shared in those all-important letters home. Neva Rae caught the spirit. "So my husband has turned out to be a thief. Well, well! And maybe that solves a problem. You know we'll start keeping house one of these days and we'll have to have furniture. So maybe I'll just leave it up to you. Then if you go to jail I'll bake you a cake with a file in it or something."

No loner was Ned, no question, but they were all homesick. "Mail call was the big time for us. We were really out in the 'boonies. Neva Rae wrote every day, my mother, brothers, often old Dr. Gentry in Roxboro. I wrote as much as I could but not daily because of missions. Sometimes several letters came in a mail call, sometimes none. Officers's mail wasn't censored, but enlisted men had to leave their letters open, and each of us did a stint censoring. Seems sad. I hate to recall that." Censored mail included flight officers, even some ATC pilots.

Meals made a safe topic for letters, if not particularly an appetizing one. "Our food was spam. Fried, baked, they probably boiled the stuff. On the table they had a bottle of Atabrine tablets which we took for malaria. The things turned us yellow, hands, skin, eyes. You had to feel a little sorry for the cooks, making do with what they had. Everything came in powder or cans. The food was better over in China, where they baked bread and we could get eggs, so we tried to eat there before we turned around and flew back to India." Like many others, Ned saw his weight plummet, his face take on a gaunt look. Better cooking was certainly something they dreamed about coming home to.

"As for feeding you when you get home, I certainly shall," Neva Rae writes back. "Do you think I want people saying I'm starving my husband? Huh? How much do you weigh now? If you don't start eating more I'm liable to catch up with you."

He didn't write much about the weather, but it was there, and no man forgot it. "It couldn't get much worse," in his words.

"There was more weather than not-weather. In the monsoons you always had problems, you pick up a lot of instrument work. You're monitoring your dials all the time. The Hump was hard on planes."

To credit General Tunner with a rule that stuck and really saved lives, it had to be his insistence on oxygen in high-altitude Hump flying.

"It was the first time we'd used oxygen on a continuous basis," Ned describes. "You wear your mask six to eight hours. It was a sealed system, on demand, with breathing activating a diaphragm. When we breathed we had oxygen. You go on oxygen at 10,000ft, that was the requirement, or earlier for night flying, as a precaution, sort of psychological. After a while I finagled a canvas cap, sort of a baseball affair with a bill and a couple hooks on it to hold my mask. Otherwise you had to put a strap around. If you wanted to talk you'd unhook one side for a couple minutes, but at 20,000 or 25,000ft, you couldn't leave it off for long. We used to look at our fingernails a lot."

Bluing nails constituted a warning sign for anoxemia, or oxygen deprivation, a silent stalker at high-altitudes and in non-pressurized airplanes. Insufficient pressurization meant less oxygen to breathe, a challenge unknown in earlier aviation. The symptomatic bluing of nails could be dealt with before dire manifestations set in—and they came fast—confusion, sweats, lightheadedness, dizziness or sleepiness, a shrinking field of vision, and at the extreme, loss of consciousness or even seizures. Obedience to the protocol paid off big-time in notching down fatalities, and pilots often fastened on their masks as they taxied out.

"Sometimes we'd be so unfortunate as to have one of those big, tough fellows who sneered at oxygen masks," writes General Tunner. "After crossing the Hump, coming down abruptly from a twenty thousand-foot altitude, his vision affected and mind beclouded by anoxia, he'd just fly right on into the ground, cargo and everybody else with him." [36]

The altitudes of the reopened southerly or lower Hump also required oxygen. Called the Easy Route, it was strangely named and plenty high, a jagged, peril-wrought terrain plagued by its own mercurial extremes of weather, hostile peoples, and other challenges. As if anyone had to remind Ned, whose longtime friend Deming had just died there. Nothing easy about the place.

Right up with anoxia lurked its evil sister, vertigo. Few Hump pilots escaped vertigo entirely, and although Ned never experienced a full-blown case, he dealt with the symptoms.

"I can remember getting a bit disoriented. We'd never experienced vertigo in training because everything was visible. It's when you can't see the ground, you may think you're flying straight and level when you're really turning, or vice versa. That's why you have the gyros there to tell you, with your artificial horizon, and you've got to believe them. You've got eighty-seven jobs to look after, and you have to do them all correctly and in a timely fashion. Otherwise things don't turn out right. Nowadays they have planes that can land on autopilot, but what we did was pretty rudimentary. You trust your dials."

* * *

No Hump run was entirely predictable, but it was always a round trip due to the dearth of fuel in China. Kunming constituted the main Chinese terminus, but Ned flew into Luliang, Chanyi, Yunnanyi, Li-chiang, and other far flung outposts, landing overburdened planes on crude, primitive airfields, onto short dirt runways that sucked sand into engines, wrecked corrosion and havoc with the mechanisms, and gummed up the works generally.

Of this enormous territory pilots flew, the Himalayas formed the northwestern wall. To the south, the Indian Ocean moated them in. East and nose ahead lay occupied China. Flyers departed the Assam Valley, radioed for weather sightings across the Hump from Fort Hertz, and kept flying on a wing and a prayer. "Turnaround to Kunming," Ned remembers, "could be eight hours or ten, depending on winds. We were cruising maybe 150mph, dealing with heavy loads and a lot of resistence. If we were lucky we had a cockpit heater that worked. They didn't always, but we had leather jackets "—sometimes layered with extra clothing beneath because cabins turned into flying deep-freezers over the mountains. Ned often waited to turn on his cockpit heater until the return leg with an empty cargo, given the dangers of stray sparks in a plane awash in high-octane fumes and leaks from 55-gallon high-octane fuel drums.

"We could take the E-route going out, that's the lower Hump, maybe flying 16,000ft to 18,000ft," Ned remembers. "The Easy Route reopened with the liberation of Myitkyina. Going east to China with your heavy load, it's harder to lift up. Coming back with empty planes we flew the Charlie route farther north. It's higher and more difficult. Nowadays planes are pressurized from 7,000ft. Without pressurization or oxygen we didn't customarily take passengers at those altitudes."

At least the engines had oxygen, thanks to those life-giving turbos. "As you go higher, you have less air and oxygen, less pressure, the air gets rarified and engines have to labor harder to operate. The turbo is essentially a fan that pushes extra air into the carburetor system. Before turbo-chargers, as you went higher the engine eventually stalled. The old gooney bird didn't have turbos. Fighters had them in some cases, the B-17 and the B-24, because anti-aircraft could go to 25,000ft, but the fighters didn't have the range to escort them until the development of the P-51. There were always tradeoffs, that was in the air.

"Then we'd be dropping down through a hole in the clouds to land in China, and most of the light below was from lanterns. Like a scene from the stone age."

How to convey the desolation of this landscape, where death called at every turn? Statistics of themselves told a terrible story. Half of all the Chinese did not survive to age thirty. Pitiless poverty, disease, brutality, and abuse were their status quo.

Drought or rainfall determined every harvest, and usually the Japanese were there to pillage first. Bare hands and rudimentary tools scratched a raw and thin subsistence from the soil.

Overflying scenes that would have looked familiar to Ghengis Khan, Hump planes threaded their course to assorted bases deep in China, along a narrow skyway perhaps 50 miles wide that could extend a thousand miles in length depending on the terminus. Logistically, the Hump was not one route but many as it fanned out east of the mountains. Some 13 bases lay on the India side, and officially included distant Karachi. Distances flown could expand or contract in response to the advancing or receding of the Japanese. Flights debarked from the rank and soggy Assam Valley at sea level and quickly climbed to arctic conditions aloft. The Brahmaputra River disappeared as the atmospheric gravy built and the slope of the Naga Hills rose ahead—a jungle readout of head-hunting tribes more or less cheek-to-jowl with Ledo Road. At the first major ridge, the desolate brown Patkai range, vegetation thinned away and jagged peaks emerged through the clouds. But even these rugged bruisers were as mewling infants compared to the high Hump. Bestrewn here and there were swaths of dense green jungle, more dreaded by flyers than the tallest mountains, and into whose impenetrable tangles, according to General Tunner, 400 Hump planes went down. Somberly he writes:

"We had lost planes all through this area, and had never again heard from many of the crews . . . Perhaps they had perished in the crash. If they had parachuted out, they may have been caught in the treetops, or injured in the fall to earth. They could have starved (or) wandered aimlessly in the dense undergrowth until they dropped with exhaustion. They could have been found by native tribes, and been mistreated, murdered, or turned over to the Japanese."[37]

Ned's ability to focus and compartmentalize wasn't always easy, but gave him an edge and he used it. What degree of fear or dread did he experience flying over these jungles?

"I don't think you dwell on things like that. In the air you're intent on monitoring your flight pattern and instruments. If you can see down, it's natural to keep an eye out for a place you can land, if you have to. I think it was flying over the mountains, more than anything—that there was always something in the back of your mind. We were all aware of planes being lost over the Hump. For the jungle we had chits in local dialects printed on silk so they wouldn't disintegrate, promising a reward to natives for returning us to a place where we could be rescued. Sure, I knew if the plane went down even my parachute wasn't going to help. I tried to be optimistic. But we were on our own, I can tell you."

Embedded deep in this savage terrain of jungles and ridges several great rivers roared through their canyons—the Chindwin, the Irrawaddy, the Salween with its great gorge within the mountains, last the Mekong, flowing into what was then French Indochina, now Vietnam. Overflown by pilot-navigation alone, the dense muffling curtain of moisture and gloom had one redeeming feature in the camouflage afforded from enemy attack. The highest point of a crossing came with the Santsung Mountains, the backbone of the Himalayas. Summits exploded from the earth as full-fledged monsters. Even along the southerly route where they could be flown lower, perhaps 16,000ft in open weather, pilots liked to allow greater margin in cloud cover, against the perils of their unappeasable peaks.

On both sides of the range water was obscene and rank, the diseases exotic, insects and varmints of assorted stripe ready to chomp into a man, aircraft technology such as it was. All this taken into account, it was not reassuring to see service crews on China ramps routinely siphoning "excess" fuel from Hump aircraft before they flew the high Charlie Route back to India. With no allowance made for the unexpected, returns could be cliffhangers for pilots as they watched their fuel gauges tremble.

What of enemy malevolence? Japanese air attacks abated

greatly as they were driven farther down Burma. The real danger they posed dwelt straight below. The red Chinese and omnifarious jungle savages were dreaded, too, but worse by far were Japan's foot soldiers, the "Nips," spread thick as tar over the terrain, uncountable across Asia, their patrols everywhere, their presence dictating the limited and dangerous air routes planes could fly. With good cause were encounters with them on the ground more feared than aloft. Pilots often felt they'd rather have taken their chances with Nazi captors than fall to the cruelty of Japanese infantrymen. In the air, too a Hump pilot could hope to dive his plane into the clouds and elude the gun sights of a Zero.

One foe never let up. Atmospheric conditions, killer weather, was the enemy that pilots got up and faced, day in and day out. The best weather—"best" being in the loosest possible sense of the word—occurs between November and March, before the arrival of the monsoons with their opaque fogs and blinding sheets of rain, 75 to 100 inches in a month to soak the air, put runways under water and periodically shut bases down entirely. From March into summer, these soups could make visibility nil almost from takeoff. It wasn't unusual for pilots to find themselves unable to distinguish the tips of their wings as they lifted into hot downpours that turned quickly to hail, ice, and snow aloft—for it was always dead of winter over the mountains—with different air aggregates ever at work to create variations in the nightmare.

South Asia hosts the most violent extremes of weather on earth, a meteorological concoction of three turbulent, converging air masses—low pressure from the westward across the mountains from Tibet to India into the Hump; the warm, wet, high pressure system from the Bay of Bengal; the frigid lows moving down from Siberia—all smashing together over the great range. Nowadays it's understood that the polar vortex, and the intense heat rising up from the Burmese jungles, intensify the clash of these diverse air masses to create a fierce, unforgiving mix of extremes in atmosphere like no other.

Yet little of this was understood at the time. Pilots knew conditions were horrific, but even the concept of a jet stream was alien. Weather was an everlasting puzzle that American meteorologists on the scene tried to predict using frontal concepts of the continental U.S., which proved useless in CBI. Reconnaissance, too, was primitive. "Jing Bao," or peasant bush telegraph, served Chennault in warning of Japanese attacks, but couldn't address air patterns brewing in the mountains or upper atmosphere. As a result, most reportage relied on pilot debriefings or radio reports along the way. Word of major storms usually arrived too late to be of use, and in all events, the sheer changeableness of conditions made it impossible to know what to anticipate. Patterns bolted across the region like juggernauts. Writes General Tunner,

"The weather on the Hump changed from minute to minute, from mile to mile."[38]

Not even bombers could climb high enough to get over the turbulence. To fly the eccentric atmosphere of an unmapped mountain range with a cargo overload of weaponry, explosives and gas, strained every nerve in a man.

"The wind stream was always precarious," Ned affirms. "It created updrafts and downdrafts. Those were what was terrible. We were flying a lot of gasoline, one of the worst things you can mess around with. The fumes were all over the plane. You smelled them in the cockpit till you got on your oxygen."

A frightening mystery rose out of the Hump experience and its weather. Based on accounts of several pilots, a monstrous unnamed summit towered somewhere north of the flight path, and crested higher than Mount Everest (29,029ft above sea level). Men blown off course by fierce storms found themselves hurled inside a harrowing updraft, and coming out on the top at 30,000ft, to confront a super-giant looming 2000 to 3000ft above their planes. The threatening pinnacle first was sighted by two Chinese CNAC pilots as they attempted to map an early route across the range. Later Hump pilots also saw it and gave credible accounts of being flung far too near the place for comfort. One of Chennault's flyers obtained permission to return in a fighter and try to photograph the peak, but was never able to find it again. That they all saw something extraordinary, no one questioned.

Did an immense mountain really rise somewhere amid countless others, as yet unmeasured or undreamt of? Were pilots blown there like mariners swept over an unsailed sea, into the unexplored reaches to a remote place where no man had looked before? Had their altimeters played them false? Or did they just lack the data to figure it out?

Perhaps not precisely any of these. Ned explains the enigma simply, in the level-headed terms of the gathering Hump experience. "With different atmospheric conditions your altimeter will vary as you move from one pressure situation to another. You're always resetting your altimeter on a flight, to reflect these differences. Sea level atmospheric pressure is 29.92 inches of mercury. As you go from one pressure situation to another, ground control will tell you the setting, and you use a knob on the dial to adjust your altimeter. You can get a false altitude reading if you don't keep changing your setting. Of course, we didn't have any way to get that information over the Hump because no stations existed in the mountains."

Plainly put, their altimeters hadn't malfunctioned, they just weren't reset for changes in pressure conditions, nor was there any way they could be over the range. So why did the story persist, despite a reasonably simple solution? Partly because weather buildups commonly went higher than Everest—a solid wall of wind and water, ice and cloud, from the deck to 40,000ft or more. Much of the range remained inaccessible and unexplored. Sophisticated modern satellite surveillance didn't exist. Who knew for certain what was out there? In fact, the mountains of the world are being measured and reassessed to this day, and even the concept of sea level still presents problems. Ned's take on the mystery reflects the common knowledge of an instrument pilot, and something of his temperament, too. Things had to make sense. He needed to keep his head and deal with the world he was in.

He had cause. They knew about inexperienced flyers sent into the Hump, recent cadets out of flight school, or mature men with minimal training in the instruments indispensable to high-altitude flying. Some had never flown the C-47 or the bigger C-46. Deployed as copilots into what was viewed as a backwater war, forgotten while their country pursued a policy of "Europe first," they took to the tallest peaks and worst flying on earth in unfamiliar aircraft, as check pilots tried to talk them through basics like raising and lowering their gears and adjusting their flaps. [39] Unprepared, unknowing, these men were to all intents murdered on the job.

Uncle Sam at length acknowledged that the Hump butchered pilots, but the truth was tardy in coming. Maybe it took that recent and infamous "wildest night" to get men selected after more rigorous training and adequate operational work.

Ned came with this newer breed. "You had to know the plane you were flying. The history of aviation went back maybe a few decades, but Hump flying was the big unknown. No one ever flew under these conditions. It was a new world."

* * *

Crystal skies over the Hump gave pilots a view none but angels could have seen before—a scene to dwarf a colossus. Rolling to the edge of the world in an infinitude of ridge on ridge, the jagged giants of the range shouldered their age-old mantles of rock, snow, and ice.

Himalaya in Sanscrit means House of Snow. Hauntingly named in their swirl of blowing white storms lay a place of heart-stopping beauty to inspire awe and blood-deep dread in men who risked plunging to their graves in its ragged folds and impenetrable ocean of crevasses.

Clean skies were not the norm. Even on days when Ned gazed out on sun-drenched peaks, invariably he'd see the familiar pile-up of cumulonimbus clouds in the distance, rising tall at the edge of heaven with winds tearing along their tops to blow them flat like an anvil. Another raging storm to brace for. He scarcely remembers a crossing in the clear from start to finish.

"You look out and see those peaks, and they're awe-inspiring. But you always know there's no place to put down."

Hump pilots trained stateside to fly the beacons found the skill mainly of use near bases of takeoff and return. Over the Rockpile they honed all their instrument expertise to a fare-thee-well, knowing even the beacons they picked up could be warped as they bumped off mountains and glanced back unpredictably as echoes. Signals also could be deliberately distorted by enemy stations. With conditions ranging from worse to worst, they all knew trouble was bound to come, but those early bad rides could be unnerving. One spring evening as Ned flew the upper Charlie route, wings icing over, a nasty squall brewing, he felt the plane lumbering deeper into the troughs, as if punch drunk finally from the merciless slugfest of winds.

From the cockpit window he saw heavy rain "falling" upwards and fast, a freakish aspect of Hump storms, and it always struck a nerve. They were getting bumped around hard, but the '46, he figured, could take the punishment. Till without warning Dumbo lurched midair, and Ned felt his body yanked in an upward slice against his seat belt. As if a trap door dropped in the sky, and in less than time to tell it, they were plunging down the elevator shaft to the next world.

Reaching for the panel, Ned flipped off his autopilot and flung his eyes across the dials. The altimeter was unwinding like a yo-yo, the airspeed needle ramping up as in the throes of a delirium. The rate of climb indicator lay dead on the dash.

They'd gone from cruising to a dark loose lump of ballast plunging down the sky, where nothing waited at the bottom but the grim reaper, into whose arms they were headed in a leaden streak, if he couldn't climb the plane out.

Pulling back the yoke to raise his elevators, he fought to get the tail down and nose up, but the tried-and-true maneuver did nothing to attenuate the wind's wild determination to ditch them in purgatory. By instinct he grabbed for the throttles and pulled back to reduce power.

Suddenly, as fast as they'd started down, a crosswind rocketed them back up. The teeth of the ferocious draft plastered him against the seat and pitched the '46 around the room like a bull terrier with a sock. He forced his yoke forward, giving it everything in his arms and torso, still nothing redressed what the vortex outside seemed bent on doing. He couldn't see out to get a bead of perspective, not the crest of a ridge, not a friendly star. It was too dark for his eyes to tell if they were up or down, and he couldn't be sure anymore by his senses. A roller-coaster of crosswinds was pushing the plane every way on the compass.

Working the controls for all he was worth, he hoped if all else failed they could brace up, ride it through, and remember eventually—what good thing to remember?—yeah, to breathe. Because without that suck of oxygen, your brain's gonna fuzz up like a woolly bear and you're worse off than before. Which meant things still could get worse. A little worse, right?

He threw a glance at his copilot sitting with knuckles frozen, eyes bugging out, mouth a taut gray fold of death. Same way he must look.

Keep the grip on, here comes the next hit.

The Commando took it in the rib cage, and as if the wind got knocked out, dropped again through the vacuum. They were going to ditch. There couldn't be any more of underneath left in creation. They'd fallen 5000ft or more, still gaining, and that was just about the rocky end for these tall parts. They absorbed another one-two punch from the port side. A low howl rumbled across the cockpit, his command to kick the power in, his copilot's shriek, what was the difference? Then from somewhere deep in the inward parts, the plane dredged hard into its store of grit and force, unfurled its wings, and lifted again.

Felt like they were galloping on all fours across the potholes of hell.

Tumbling along on the berserk runaway winds, they fought their way west trench by trench. By hairbreadths the altimeter needle nudged up. The RPMs ground hard and steadier. Nearing the edge of the ridge he set up their descent. Left hand gripping the yoke, the shirt under his leather jacket soaked cold to his skin, he tuned in for landing instructions and heard the voice of sanity issue calmly from Dinjan tower. Minimums fair below. By now the updrafts and downdrafts were breaking camp as brusquely as they'd ambushed earlier. Through a breach in the clouds Ned guided Dumbo down, circled, and gentled his ample girth into the field. The jeep was there to ride them to Operations. He filled out his paperwork while his copilot trailed along in silence—hadn't spoken since the start of this carnival ride. They turned to head for the bashas. Ned wasn't sure the fellow had breathed yet when he uttered his first words.

"Couple times there, I thought we were goners."

Ned slapped him lightly across the back. "Hey, no big deal."

They were alive. That was the big deal.

Every Hump pilot enacted this scene in multifarious variations, and it invariably came. Writes General William Tunner, "Planes could drop at the rate of five thousand feet a minute, then suddenly be whisked upward, at the same speed."[40] On piston aircraft the noise alone could strip away the last of a man's mettle. Ned remembered his piston rods and cylinders, if not the entire engine "out there banging away. I'm surprised they did as well as they did." Still, the planes performed somehow. Pilots did the well-nigh impossible. If all else failed, if they couldn't recover, they bailed out if they could, went down with their air-craft if they couldn't. This was the gig in the Assam. Planes bore up to the punishment, usually. Men bore up.

"Autopilot is very capable in a variety of flying conditions," Ned reflects calmly, as if Hump aircraft fluttered along the ridges, sweetly as moths. "It was a great advance, a godsend for pilots. You could set a climb, and you didn't have to manhandle it all the way up. Autopilot can take a plane through and leave pilots to monitor the dials, watch for ice build-up, and tend to the engines, gauges, radio and navigating. Seems backwards now in an age when computers can do so much, even land a plane. You didn't leave your seat back then, or take a break. If you got into turbu-lence, you'd switch to instruments and try to compensate, but coming off autopilot in these times is sort of a last resort. The whole plane's thrown around in the downdrafts and updrafts, no matter what you tell it. If you're losing altitude rapidly you want to reduce power, but it may not help. You can be in an updraft one moment and a downdraft the next . If you're going up and losing speed, you may try to correct that by pushing forward on the throttle, or if you're gaining you're pulling back.

"We got bounced around pretty heavily. You couldn't climb over it like today. We had some anxious moments. I can tell you, pilots don't like turbulence any better than anyone else."

* * *

If Claire Lee Chennault's strength lay in his bent to call things like he saw them, the same quality may have been his greatest handicap back in the days when the tradition-laden infantry rele-gated air tactics to the outlook of the Great-War, to tilting and face-offs where flyers lined up in the sky and duked it out like dandies. Invited by Madame Chiang Kai-shek to assess the state of the Chinese Nationalist Air Force in 1937, Chennault was appalled by the shabby aircraft and facilities, and the low level of training for men who went down with crippled planes rather than

lose face and bail. Here at last, his plain talk had got a hearing.
Impressed by his forthrightness, the Generalissimo asked him to
remain and rebuild the force.

The stage was set for a legend. The lanky middle-aged Texan
who'd honed his flying skills barn-storming back home, put out
the call for his Fourteenth Volunteer Bombardment Squadron,
and started training pilots his way. By the time Roosevelt signed
his secret pre-war order allowing Americans to fly in China, and
not lose their military status back home, Chennault had shaped
up a fighting force. His President's action, its rationale to have
air power ready "when war came," effectively landed the U.S. in
World War II before a formal declaration. Chennault's recruits
signed one-year contracts, they drew handsome paychecks compared to their army salaries, and had a bonus, too, for every
enemy plane they shot down. There it was, a year before Pearl
Harbor—U.S. pilots were in China through the work of
Chennault's Fourteenth Volunteer Group, aka. the Flying Tigers.

Four months after Pearl Harbor, Chennault officially merged
his group into the China Air Task Force, or CATF. Now, Chinese
and American servicemen were sanctioned to fly together.
The Chinese Nationalists made him a brigadier; in due course,
he was given a star by the U.S. military as well, when CATF was
made part of the U.S. Tenth Air Force. Having 12 bombers and
50 fighters at his disposal—many being derelict P-40s he'd commandeered by sundry nefarious means out of Rangoon and elsewhere—Chennault's mission with the Tenth was to provide air
support for China, and defend their nascent struggling lifeline
over the Hump.

Why did the U.S. want Chennault in the Tenth? Bluntly put,
to gain ascendency over a maverick who never did the expected.
Chennault's first principle, one he instilled in his pilots till it
became second nature, was never to fight the enemy on the
enemy's terms. Move in like lightning for the strike, dive, risk a
shot only if the target flies directly in range. Hardly the old gen-

teel approach, it proved spectacularly successful in keeping the foe off balance and their casualties high, despite the numerical and technological superiority of their Zeros.

Chennault's unorthodoxy, his drive and ability to think out of the box, let him deliver where others saw obstacles and impossibilities. If only he could've been given rein for an all-out fight by air, and been allowed to hit the Japanese every place it hurt. But as was already manifest, success didn't exempt him from ruinous rivalries. Stilwell fought him at every turn; the disposal of Burma brought their conflict to a head. Chennault's proposed air offensive, whereby he promised to disable Japan's army and navy strongholds in China, disrupt their supply lines, destroy their morale, and drive their backs to the wall, asked little enough of Uncle Sam—100 new P-51s, and 30 B-25 bombers, all told. Implicit was that Vinegar Joe's Burma campaign wouldn't be needed, and thus much human life and suffering would be spared.

Huffily Stilwell brushed aside the request, reminded Chennault that CATF's mission was to defend the Hump, and angrily demanded an apology. Fat chance he'd get one.

Opportunely for Chennault, Wendell Willkie planned a trip to Asia. As ambassador without portfolio for the President, the celebrity/politician (who opposed FDR on the Republican ticket in1940) was traveling the world, talking to leaders and common folk alike, and generally snooping into things. Reaching China, and with Madame Chiang's blessing, he met up with Chennault. Profoundly struck by the general's conviction that he could inflict so much damage on the Japanese by air with so few resources, and convinced a larger air offensive was the only way to move the war along, Willke urged Chennault to write President Roosevelt directly, put his disagreement with Stilwell on the table, and ask for the additional fighters he wanted, along with 30 medium and 12 heavy bombers. Willke himself carried the missive back to the States and placed it in the hands of FDR.

Never one to beat around the bush, Chennault writes, "I can

cause the collapse of Japan, and I can make the Chinese lasting friends with the United States for generations." He recalls to FDR the past successes of his volunteers and the trust he'd earned with China's leaders. He denounces, too, the rigid thinking of commanders who didn't understand aerial warfare in China. "It is essential that I be given complete freedom of fighting action so as to be able to deal directly with the Generalissimo and the Chinese forces." [41]

Stilwell, brimful of jealousy, wouldn't give an inch. Chennault's strong card was that he'd proven time and again he was as good as his word. Given their differences, important questions arise: How much life in Asia, for allies and foe alike, might have been spared had Chennault been allowed to pursue his air offensive, in lieu of Stilwell's costly land wars? How much good will might America have gained, had Chiang Kai-shek been freed of the Japanese, to put his forces into beating back the Communists and rebuilding his nation? Might the China-Formosa split have been avoided? And the ultimate question: Had Claire Chennault's planes eviscerated the enemy, could humanity have been spared the horror of atomic warfare over Hiroshima and Nagasaki?

For the rest of it, had FDR and his chiefs believed Chennault was bluffing, might they have given him the small number of planes he asked for, and been none the worse?

Thanks to Stilwell, CATF ran so short of supplies and fuel that combat functions had to be discontinued. The disaster raised the alarm. Generals Arnold and Marshall, and Ambassador Soong protested in Washington. Keeping Chennault tied to Stilwell's strings had turned into a debacle. In a dramatic turn-around, the U.S. abruptly promoted Chennault to the rank of major general, again over Vinegar Joe's protests that he'd been "sufficiently rewarded" already. Brooding and embittered, Stilwell deployed his men into Burma around the time the Fourteenth Air Force reactivated in China, in March of 1943.

Old Leather Face had managed to stay in the game, but still remained far too under-supplied to take an air offensive. Stilwell, he argued vociferously, was squandering the whole Hump Airlift in his campaign for Myitkyina.

Chennault was no different from other CBI commanders by his dependence on Hump pilots for his war needs, all of which evolved into another conflict after Stilwell's departure, this time with General William Tunner. Chennault started diverting Hump pilots to missions inside China, thus tying up planes and wrecking havoc with those clockwork schedules. Irately Tunner told him to find his own fleet of C-46s and crews for the China side. "When I proposed it, General Chennault screamed bloody murder. You would have thought I was reaching into his command and taking his planes away from him, instead of merely operating my own." [42]

Such were the rivalries of command in Asia. The huge region had languished in a holding pattern for years, an arena of turf wars for generals and of problems endemic to a part of the world where nothing could be addressed short of a no-holds-barred commitment. Tunner won his round with Chennault, but in other ways, all sides lost.

Nothing less than defeat for Germany, and Armistice in Europe, would create a shift favoring Asia in U.S. priorities. Uncle Sam got serious about moving supplies, airplanes, equipment and fuel once intended for other theaters, into India for transport to China. It was a massive effort, clumsy, overwhelming, well intentioned. On the India side, an avalanche of supplies landed in Karachi and Calcutta, then had to reach the Assam Valley via an antiquated railway system, which though improved under military oversight—slack speeds increased, cars added—formed a pathetic link in the chain, as were the small ferries and barges that ran into innumerable bottlenecks as they plied the Brahmaputra River. Ledo and Burma Roads both remained susceptible to ambush, and even in the best times, their single-file, serpentine trek never succeeded in moving air-craft, tanks, fuel and men in quantity to wage a world war. Only in theory were the Chinese in better shape by overland access to Myitkyina.

In reality, the Airlift accomplished what no other route could do. Hump pilots on day and night missions ferried more supplies and personnel over the Himalayas than Burma and Ledo Roads combined. Watching his charts, General Tunner proudly noted that the operation kept breaking every tonnage record on the books.

As for Claire Chennault who had held to his watch for nine years and seen his bitter enemy Stilwell recalled to the U.S., it was presumed by now that he stood next in line to command China. British Supreme Commander Admiral Lord Louis Mountbatten, in anticipation of the promotion, designated more bases in India to fly supplies, and had new airfields scoped out near Myitkyina. By May of 1945, the Tenth Air Force was mobilizing to fly under Chennault.

* * *

No aircraft in all World War II ever took to the Hump like the Curtiss-Wright C-46. Fundamental, pleasing to the eye of few, the bulky plane earned the loyalty of pilots when they saw it bear up under adversity and go the distance. That said, General Tunner, hated the "bug-ridden" thing, as scathingly he describes.

"The C-46s, at least the early ones, were particularly prone to engine failure. At its full load, this was a cumbersome beast that could cause anguish when it lost an engine. The wing would automatically drop (and) the only thing you could do would be to put full power on the one remaining and pray it would hold up. Even full power on the remaining engine couldn't save you. The ship would stall, and a black spot at the end of the runway would be your memorial. It took the rugged conditions of the Hump to bring out the worst of Ol' Dumbo. (It) was never a completely dependable plane. C-46s were killing crews."[43]

Tunner considered the smaller C-47 superior, but the plane he really favored was the spacious C-54, though neither it nor Gooney ever really handled the high altitudes where Old Dumbo climbed through his paces. True, the aircraft's delinquent reputation had to undergo rehabilitation, but by 1945, pilots were flying it proudly as the Commando. This was the plane that mainly kept China alive and fed the war of the Orient.

The physical configuration of the '46 met the mark. The cockpit set 21ft above the tarmac, dwarfing the '47 or any fighters arrayed below, even some of the bombers. The aircraft had two bays, an upper for troops or cargo, a smaller one below to house the auxiliary power unit (APU) and provide a stash for such overflow as might prove useful. Reached by a trap door in the cockpit, or a small exterior door from which pilots could bail out below the aircraft, the low space had Ned on all fours from time to time, and provided a repository for the weaponry of the unpredictable half-wild troops he'd soon be flying.

The Commando's props were powerhouses compared to Gooney's three-blade, 1250hp hydraulic propellers, which if they windmilled, the only solution was to feather them, hope the oil in the hubs came out steady, and try to limp along home.

"On the C-46," Ned explains, "if you have a runaway prop and lose your RPM control you manage it electrically using a switch on the pedestal to bring the RPMs back till you get the plane in for emergency landing. If all else fails, you still can feather but you've got more control. In an engine failure you have to feather the prop anyway, but for a propeller failure you can use your electricity." The propellers on the '46 had four mammoth blades and 2000 horses to power each engine. Turbos injected air into the carburetors to raise the ceiling. The APU in the lower hold was there for extra battery power. Lower side windows in the cockpit improved visibility. One advance simply involved stripping the olive-drab camouflage from older planes to the bare aluminum, for even the weight of paint created deadly drag.

Ned's observation that flying conditions in the Hump couldn't get much worse would have been echoed without exception by any pilot who flew there. To get thrown all over the sky in winds and storms was hard enough, but other unsettling phenomena had to be dealt with, one being St. Elmo's fire. Occasional reports had it melting off the antennas of aircraft. Whether damage was sustained or not, its occurrence made for an unnerving experience. Ned explains.

"You look out and see your propellor with this big ring of fire around it. Or there's a ball of fire tracking straight up that wing towards you. Or else moving up and down your wing. There you are thinking about what you're hauling, which could be ammo or high-test aviation gas. Then that fireball comes clear up the wing and starts to dance around outside the cockpit. Till suddenly it disintegrates with a loud exploding bang."

So there was fire. And there was ice, all myriad forms from sleet, hailstones, snow and crystals, to fierce storms containing frozen mixtures of every configuration. Ice glazed interior and exterior cockpit windows, and could build up on the front of windshields in sheets several inches thick, weighing down already overloaded planes and making them sky-borne icebergs. The weight of ice interfered with their capacity to hold altitude. Multiple fatalities in the C-46 over the Himalayas were attributed to carburetor ice alone, though with warning, pilots could combat it. Men spoke of flying in a turbulent icy hell that beggared words.

"We had a lot of wind, a lot of weather, a lot of turbulence, a lot of anxiety," Ned admits. "You run into icing conditions in very bad weather. Coming out of the tropical climate in India, climbing into frigid conditions, the propellers would start throwing heavy chunks of ice against the fuselage. It's hard to imagine how ice can build up on something that's going around at more than 2,000rpm's. We'd hear it breaking off, crashing against the sides of the plane. It was loud coming off at that speed. POW! We could hear it up there over the roar of the engines."

Pilots regulated a tube from inside the cockpit that fed a glycol deicer into the hub of the props, but in heavy storms ice prevailed to varying degrees. The vertical and horizontal stabilizers also had deicer boots which could be activated from the controls. These expanded or contracted, moving back and forth in a slow dance to dislodge ice, but oftentimes the deicers got so severely cut up by the rocks on Chinese runways as to be past use. Of course, cockpit heaters didn't always work either, so it could be a cool ride all round.

Close shaves with lightning were commonplace. In conditions of low light, a flash could blind a pilot unless he compensated by turning up the rheostat to brighten his panel and offset the glare. The mere proximity of the terrible jagged streaks, each carrying enough electrical charge to light a city, was to feel the encounter of a sky-borne killer.

Gyros went crazy in the turbulence, and may have been the source of a few glory-tales of Hump flights solely on needle and ball, although needle-and-ball flying was all but impossible under such conditions without something near divine intervention.

"I didn't know of any pilot who could've flown needle and ball alone," Ned states. "It can be done, but it's terribly difficult. We practiced it as cadets on our Link trainers. If you lose your artificial horizon, you've got the needle and ball, but the artificial horizon is a real advancement. The radio compass is just for navigation, the instrument that directs you to a station. I never had a total instrument failure. They're very reliable actually."

An optimistic statement, as mostly his were.

The seeming simplicity of the artificial horizon, or gyro horizon, makes it no less a breakthrough, a miracle dial with a small "airplane" that moves above the center line when climbing, to one side or the other if turning. The dial was replicated on the pilot's panels, along with its close companion the gyro compass, which together were used to establish the setting of the autopilot. The magnetic compass crowned the center panel, and could be used to reset the gyro if it got incapacitated in bad conditions, or as Ned says, "as soon as you get your ship back on a reasonable plane," as one ever-earnestly hoped to do.

The bird-doggin' pilot's companion, the radio compass (ADF) notoriously failed over the Hump, frequently behaving like the hand of a clock cut loose on its spring owing to the constant electrical interference of high-altitude storms. Ned certainly remembered.

"I've often sat and watched that bugger spin. All you can do is wait till the electrical interference dissipates, or you get closer to the station. Navigation was pretty rudimentary. If you lost your radio and compass, then you'd try to fly time and distance."

This is what happened one morning following a run into Yunnanyi. Or was it Chanyi? Runway too short. Can't see the start or end of it.Very late he'd dropped his plane through the nebulous darkness on instruments, faith, and a few glimmers from village lanterns, appearing like fireflies against the vast Asia night. He'd talked his way down till his wheels screeched against the heavy steel matting of the airstrip, which for what it's worth, beat conditions hands down over his previous flight to Chihkiang, where you landed in dust so thick a hurricane couldn't have blown it back out of your props. Not to neglect your ears and windshield. Dust in Chihkiang rendered ground visibility nearly as bad as the high Hump in the clouds. But never mind. Flight accomplished, he'd put his feet up for an hour, and turned around.

Now he's on the way back, flying the high Charlie route, weather rough, visibility zero, conditions grim all told, and to top it off, he's lost his signal.

"All I could get was a lot of static. I couldn't do a radio check, and of course, I had no land fixes. I'm trying to work this out, when I get just a bit of a signal, very faint, but a long time before I expect it. Which if accurate, tells me I've got a 100 knot tailwind."

What if not accurate? Signals ricocheted all kinds of ways in the Hump. He knew, too, that pilots had been thrown off track by deceptive beacons transmitted by the enemy, and then killed in high altitude crashes. The Japanese could set up shop, unleash their mischief, and move on. For myriad causes flyers dreaded them more on the ground than in the air, not least because on the ground they moved largely at will, towing along their diabolical radio equipment.

What to do? Given a hundred knot tailwind and the pummeling they were taking, he had to think fast. Had they been pushed farther than he thought? If he set up his descent based on a malevolent signal, he'd wipe out against one of those heinous pinnacles, with hardly a blot for a memorial. He sure couldn't trust his senses for a reliable measure of what was happening.

With his panel crippled, sailing the high lost currents above the range, no outer view to offer anything other than a blind twilight from which the occasional rock face emerged, he was down to his last resources—not to be underestimated—careful training and calm thinking.

"My instruments aren't much good in this case, and I can't see anything outside the plane. So what I do is put my wind and direction into my E6B, do some computing. Right there I know if my computation's right, and that's a big if, I've got to be much farther along than I'd thought.

"I'm between two devils here. I decide to set up my descent early, knowing if I'm wrong, I'm going down in the mountains and hitting a big rock somewhere. But with a 100mph tailwind, if I don't come down, I'll overfly my base. When I break out at 500ft, it's daytime— still raining hard, but there's the Brahmaputra River below. That's our marker, and I was glad to see it. Home base and Dinjan field ahead."

He'd trusted his assessment. Confirmed, too, that when a pilot's good friendly RDF fails, he still had this fallback position. Primitive compared to his panel, still the handheld E6B with its slide-rule and round-faced dial arrangement, could be put to work to get an aircraft out of assorted jams. With his E6B, a pilot could work with ETA (estimated time of arrival), ground speed, rate of drift, estimated wind speed, and so on. For example, when it took him longer than 2 hours to go 200 miles (as happened in flying against such horrific atmospheric forces) he was therefore slogging against headwinds at an airspeed less than 100mph. He then could compute the speed of those ferocious winds, reassess his ETA, and make educated decisions as to how to proceed. A pilot always wants a Plan B. He keeps his head.

By staying positive, Ned met challenges he refused to glamorize as war-lore, though he and every pilot there became part of that lore. Using what he had, and refusing to panic, took some of the edge off difficult flying. Instruments failure near your airfield? You fly visually as soon as you can, manipulate your flaps, time getting your gears down. Lose your hydraulic pressure? You crank the gear down manually, or failing that, establish your plane for a glide, taking into account your principal areas of drag. Carburetors ice-up in the engines? A pilot knows how to apply heat from the cockpit.

The lessons in the mechanics and structure of aircraft proved every mote of their life-giving worth over the Hump. The carburetor, for an elementary example, is a Venturi tube, narrow at center, widened at the ends, that controls the flow of fluid so it comes through the center at great velocity to create pressure to mix fuel and air as they go into the spark plugs, which in turn, furnish the electrical charge or "explosion," of power to turn the cam shaft. Which connects thereto with the motor, the wheels, and so on. The ability to see and organize in his mind the interlocking, interior workings of his plane, this foundational knowledge, helped Ned to think under trying conditions. Was he scared?

Of course not. Wasn't in his vocabulary. Maybe another word.

"We were apprehensive. To be scared or frightened, you lose some of your ability to reason and act. But certainly, with the conditions we were in, you can get apprehensive."

He cracks a smile. "I'm not saying you're not on the border of being scared."

The instruments he and other pilots used with reasonable confidence seem crude by modern standards. Many are obsolete but they represented breakthroughs in navigation. The needle and ball dated from the early 1920s, the artificial horizon, its infinite superior, came in the late '20s. The E6B, sort of a circular slide rule—they called it the Whiz Wheel—could calculate fuel consumption per hour, wind correction, air and ground speed, ETA, and other factors that helped a pilot assess where he was and when he'd reach someplace else. Constructed of cardboard or aluminum and celluloid, the logarithmic device with its dial could compute assorted ratios and other problems. Pilots used them in training, probably expected to have little need of them thereafter, but in the difficult flying of the Hump, the device could be a lifesaver.

A plane's antenna (meltdown being a rare event despite lightning) often failed to pick up beacons over the Himalayas due to the terrible storms and extremes of weather interference. If his radio compass or RDF started to fade, a pilot rotated his antenna in hopes to pick up other broadcasts or frequencies, which could be different control towers or even commercial stations. His radio compass provided means for working off a particular beacon. The needle points to the station. For a pilot to proceed to that station, he follows the needle (hopefully, but not always, at zero). A directional compass on the other hand, points north (still hopefully) whether the plane turns east, west, north or south. Using the artificial horizon, his gyro compass, calculations to compensate for headwinds, tailwinds, crosswinds, and those killer up-and-down drafts, a pilot hewed to his course, tried not to get disoriented, tried to make corrections.

They all lived with reports of men who went down, of aircraft that broke up midair or disappeared without a trace, or couldn't be kept right side up by flyers fighting for blessed life at the controls. Whether news came from their own squadron or another, it was sobering. Ned's friend Deming had been flying a C-109 (a converted Liberator) commandeered for fuel transport. These "flying bombs" with their labyrinthine problems and mechanical flukes were notorious for crashing on takeoff. Deming's plane, however, exploded mid-air coming out of the lower Hump, and was located by native witnesses. Called with gallows humor the C-One-O-Boom, the '109 (like its cousin the C-87) was one of several default planes impressed into transport. Whether the things burst into flames in the air or at the end of a runway, chances were virtually nil to get crew out alive. This meant death for another pilot, copilot, navigator, engineer, crew chief, and whoever else were there to pray and pitch in at the emergencies that struck.

Hump pilots strove for the extra measure for safety. One lapse could mean the end. Their aerial maps being rudimentary, they tried to pick up on a radio frequency, identify the Morse code signal, and home in. In less stressful times, such as they were, they might search out the Army Air Force station, or Tokyo Rose who played all the latest music from the States. (No one could figure out how she got American records faster then the Armed Forces station at Chabua.) For those interested in such things, the ADF ranged from 300 to over 1500 megacycles. Rose showed up at different frequencies, always to play the best hits from the good old U.S.A.

Ned cracks a smile. "She was sarcastic about Americans and our intervention. Told us we'd never make it back to the U.S. That the Japanese would shoot all the Hump pilots. Tried to feed us stories about what our wives or girlfriends were doing with the 4Fs. Her voice sounded very American, good and modulated, no accent at all. She'd say, Hello, GIs, then start her diatribe. But usually she was pleasant. You don't undermine morale by getting angry, you try 'sincerity.' She was a joke with us, but we liked the music. She had Sinatra, Benny Goodman, Tommy Dorsey, Bing

Crosby, Kate Smith, the Andrews sisters. We heard them all. And Vera Lynn—they called her the soldiers' sweetheart."

Tokyo Rose may have been a joke to the men. More unnerving without a doubt were her occasional broadcasts containing the names, ranks, and flight destinations of Hump pilots.

As the thick soupy monsoon rains of May, 1945, gave way to still more soupy monsoons in June, the men flew, the war raged on. Then came a letter from home that Ned had to read again for his reasoning mind to take in. He and Neva Rae would be having a little baby around Christmastime. She'd gone home to East Prairie, and been so ill with pneumonia that when she got well, she didn't realize for some while that she was expecting. She and her mother had just set to painting the kitchen together, suddenly this wonderful news, and she was busy as could be counting the days till he came home, and the baby arrived. They were making a bassinet with ruffles all round, her parents drove her to doctor appointments over in Cape Girardeau, she was sewing the livelong day, soft baby clothes of cotton and wool and flannelette.

How to get a fix on this. He was going to be a daddy. He sat down on the end of his cot and pondered. His E6B sure wasn't going to help him now.

His reply had to be right. What to say?

Well, he was mighty glad. Hoped she was doing fine and taking care of herself. Good to know her family was there to help. Only thing puzzled him—he needed to understand the workings of the matter—sort of a technical question. He saved it for near the closing.

"What," he asked solemnly, "is a flannelette?"

CHAPTER 6

"AN EPIC OF THE WAR"

SWINGING THROUGH THE SCREEN DOOR OF THE MESS HALL AT six-fifteen in the morning, Ned caught a familiar face in the early herd of passing men. A friend. Acting like a stranger.

"That you, Mayer?"

"Ned Thomas, as I live and breathe. What's up?"

"Same routine. Eat that fried spam and canned grease. Wash it down with an Atabrine, and head out."

"Well, we stay busy. Haven't seen much of you, 'guess because we got different bashas."

"Guess so." Ned merged into his friend's stride. "Say, Jack, if you need anything—"

"Not a thing. But now that you mention it, I gotta tell you, I hear some wild stories. Kinda gets me thinking. And what I think is they're ninety percent bull."

A cautionary twitch flicked the side of Ned's mouth.

"Take search and rescue." Jack wheeled around to face his companion. "And don't laugh this off, Thomas, they're playing us. Say you've yelled your last May-Day. You're going down. Truth is, it's no different if you pitch into the Rockpile or the jungle. You stand up, find a pulse on your arm, and figure you're lucky. Well, don't waste your time, my friend. Because where rescue's concerned, we are less than half a needle in a haystack the size of Saturn. So just grab yourself a handful of grass—"

"Aw, Jack, knock it off."

"Weave yourself an itty-bitty grassy skirt"—Jack's hands dangled an invisible garment somewhere near his midsection—"find yourself a woman, and make up your mind to go native. Because you are never coming outta there again in this world. What else do you think happens to those guys they never find?"

Chuckling, then shaking his head, Ned strode into Operations. He and Jack Mayer went back. Moody for Advanced. Malden and Austin. Flew the Fireball together. Is Jack really worried about going missing, or just entranced with the wild scenario? He heard the guy hasn't checked out, still can't captain his own plane, and it's July. Maybe got into something bad up there and lost his head. Hard to figure, though. Jack's a good pilot. Maybe his mouth got him into the soup. Well, if you ditch with Jack Mayer, chances are, you'll go down laughing.

He runs an eye over the roster. Checks the weather reports from incoming pilots. Rain, wind, ice. So what's new? Scans his flight plan to Kunming, a place he can wing to blind, and typically does, on instruments. Flight plans are time-centered. Granted you never know how long you'll wait while they stagger takeoffs to bring in the returning planes. The wind can blow you off course, too. Still, if you don't arrive in timely frame, they'll know your general route and notify search and rescue. Preferably before you gotta go native. He pushes out the door and heads for the ramp

His ears roar in the din of the planes idling for takeoff, as many more in the stack waiting to land. He'd felt a downpour brewing from the minute their enlisted CQ charged in with his flashlight to rouse the men who'd be flying. The damp seeped through mosquito nets, stuck to your skin while you dressed, hung thick in every breath you drew. If tower manages to shuffle out the present group fast, he might be over the fence and climbing a couple minutes before going on instruments. But if water blots out visibility, there'll be a stifling wait in the cockpit for a break to see the runway again. You don't do takeoff on instruments. Got to have your minimums.

So who's his copilot? Brewster, can't place the guy. A lot of times you don't know the men you're flying with anyway.

Five minutes later he's walking the circle around his C-46, flinging a hawk eye over struts, flaps, the long wide blades of the propellers. He squints through the half-twilight of the overcast for details like fuel oil leaks. Mechanics supposedly watch for these things, but any prudent pilot's going to do it for himself. You never know if somebody maybe ran a truck into a wheel and didn't want to report it. Plus, the rock runways in China are holy murder for cutting and blowing out and generally slashing airplane tires to ribbons.

Inside the aircraft he casts the same trenchant eye over his load and tie-downs, dominated today by the sprawl of two jeeps, which on inspection appear securely cabled. There's 100-octane aviation fuel, several 55-gallon drums reeking in the humid air. Fuel plus two jeeps? Come on. The vehicles hog the space, but he takes note of some crates which could contain armament, rations, airplane parts, apparel, tennis shoes—prized footwear of ragged Chinese soldiers—also a large box bearing a red cross for medical supplies. This ship is for sure overloaded.

So what's he going to do, weigh it?

Still, things look evenly spaced. If a load master knows how to pack a plane, he'll balance it right. If he doesn't, it's going to be hard to control and fly. Too much weight forward, it'll keep the nose heavy. Too much aft, and when you're trying to take off, that tail's going to be dragging a long time. Whatever he's hauling, he wants it insulated against sparks and tied down hard. Because when you hit those thunder-bumpers, you don't want it all coming at you through the cockpit wall. He tosses his parachute someplace near the boxes and reaches the cockpit as droplets of rain start to speckle the wide front window.

Coming on seven in a midsummer morning, it's dark as Russia outside. He's in his seat going through pre-check, starting in the middle with the pedestal behind the throttles. Crossover valve down, emergency brake down, he moves up the list to carburetor heat—cold—landing gear down, tail wheel locked, wing flaps up. Propeller feather switches normal, aileron tabs ditto, elevator tabs, rudder tabs, gas in the gauges, oxygen in the system, check and check. Deicer valves off, fuel selector for correct tanks, on and on. A hundred on-and-on's. At his right, Lieutenant Brewster cinches his belt, surveys the controls, then pulls out a handkerchief, blots a sweaty forehead, and wipes a porthole in the foggy glass at the side near his face.

Using the magnetic compass, Ned sets his gyro compass and adjusts the artificial horizon for eye contact. Everything on point. In training they could check in with the compass rose—technical reassurance for flyers that their compasses read correctly, even if the "rose" were only an arrow on the ground or a chalked circle at the revetment. Since your compass floats in liquid, most need checking once in a while. Stateside ground crews could jack up the rear of planes to level them, and make certain the compasses pointed to magnetic north. Here they just sometimes passed information to the pilot that the compass registered so many degrees off zero so he could compensate. Say you're 3 degrees off, you want to head 45 degrees northeast, you head 48 instead. Not that Dinjan field, India, is exactly abloom with compass roses.

Not that anyone likely could find one under all the water anyway. Not that they had means to go round hoisting up the rear ends of planes as big and heavy as the '46, like so many puppies getting distemper shots. Not that compasses work all the time over the Rockpile. Because they don't. Not that he's going to worry about it now.

They're going to be on instruments all day, sure as frogs hop. And if the weather turns bad—*if* expressed in some utterly ludicrous sense—then his magnetic compass is just a round thing with fluid that's gonna bounce like a ball dribbled around the court. So you set the gyro compass by it while it's stable. Everything checks, with a double check on brakes because when the propellers start to turn, the plane's gonna jump to life and want to move. He starts his left engine, hears the power crank in, then the right. The instruments light up on the panel, things are humming. He radios the tower, and revs the engines.

"Army 8209 Dog, ready for takeoff instructions."

"8209, hold position till further, over."

"8209 Dog. Roger." He turns to Lieutenant Brewster.

"Been here long?"

"Came week before last." Brewster lifts his garrison cap to wipe more shine from the thinning patch at the top of his head. "Had a couple months in Karachi, flying some of the brass between bases."

"You'll do all right," Ned says evenly. Voice from the tower cuts in.

"8209, clear to taxi to run-up position."

"8209. Roger." He steers the weighty plane from the ramp and down to the turn. It's so dark out, looks more like night coming on than a fresh morning. He pulls into position, spacing up his aircraft with the others. Repeats the cockpit check meticulously, bottom to top. Engines, magnetos that supply their electrical power, a 200 mag drop, good and good. Locks the tail wheel to keep the plane in line. Reports to the tower, gets word to hold his position, and it looks like they're coming up fast.

"Tower to Army 8209. Proceed to runway 18."

The words tell him to move 180 degrees to one end of the runway. Runway 36 would refer to 360 degrees and the other end.

He hears, "8209 clear to move to takeoff position."

Then, "8209, you are cleared for takeoff."

"Roger. 8209 rolling." He revs both engines to a roar. The

Commando lurches down the tarmac, its lumbering, pitching stride not lacking in a rough sort of swagger. Hugely ungraceful of gait, still there's a rugged bravado and authority about the big plane on the ground. Accustomed to the rocky ride, Ned advances the throttles, and using the rudder to steer, licks up the runway at takeoff speed, releases the brakes, and tweaks in some forward trim. Tail up, he pulls back the yoke, feels the weight of his body shoved against the seat as they break ground and lift. In the same instant, the sky cracks open like a rotten egg and unleashes an ink-gray innundation of cloud and water that buries Dinjan field from sight below.

He thumb-signals from the throttle for his copilot to get the gears up, and pulls back the air speed. It's a fight at the yoke to lift a '46 with a load against the wind, but you don't stay at take-off power. Gotta crank in a lot of trim to hold the plane in a climb. He adjusts the trim tab settings that govern the movable far-edges of the vertical and horizontal rudders—the marvels of flying. In the path ahead, the ridges of the E-Route begin their steep shamble into marching order. They're pushing 120mp. You do some dicey flying in a climb. Even over the lower Hump, it's time and work getting to altitude, and it's altitude and speed a pilot wants, his best friends. Attain both and you're in better shape.

By and by, the next thumb-up. They're going on oxygen, and there's Brewster over in his seat, still sopping around, this time his trouser front with one hand, as he tries to plug a rivulet in a corner of the window above with an index finger of the other.

"Confounded river in here," he mutters.

"Get used to it," Ned comments cheerfully. "And get out your mask. Pull on your jacket if you're cold."

"How often do you guys fly soaked?" His companion reaches around for his sleeves.

"Lot of times I take off with an aerial map across my knees to help catch what comes through the windshield. You can't seal up these planes, what with windows that open and sealants deteriorating. When you get in the monsoons, that water is for sure coming in. Not that I'm saying we got Niagra Falls in the cockpit."

"You don't have to say it. Anyone with eyes in his head, Capt'n, he sees it plain."

They're into their masks, plane trimmed, everything cleaned up, they're cruising. The cockpit chills down to a dry berg of ice. Their feet smart, then go numb in their high brogans. Ned pulls his leather flight jacket from behind his seat, and decides to switch on the heater. Moving against a steady headwind, they monitor the panel, look out for ice, check engine performance, watch the gauges for altitude, airspeed, cylinder-head temperature, and fuel. The surround is an unseeable, stygian dusk of hail and hoarfrost, eminently normal conditions as they hit the turbulence. Bit of a rough ride now. The '46 leapfrogs through it as they navigate their course. Unseen below, roll the Patkai range, the steaming jungles, tumultuous rivers in their canyons, the ferocious storms of the Santsung spine, every manner of weather up and down to keep a pilot centered on flying. Always things to tend to. Four hours and a half, and they establish their descent. Good time, all told.

"Kunming tower, Army 8209 Dog. Fifteen minutes west of station at 16,000ft. Request landing instructions. Over."

"Army 8209, landing runway 27 east. Winds 260 degrees at 20 knots. Altimeter setting thirty-point-one-two, ceiling1200ft." The information tells him the wind is east-northeast at 20, and what angle to approach the runway to land into the wind. You always take off and land into the wind, specially on that downwind leg, because a tailwind can push a plane a lot farther than where you want to put it. Tower radios again, and he's number four in the stack.

Too banked in to see the other planes, he proceeds east on the beacons, flies the checkpoints in descending increments, breaks out, and he's on the approach. Tali Mountain with its old

Buddhist monastery, its broad calm lake and treacherous jagged-peaked surround, fit themselves together in the field of vision like artful touches in a Chinese brush painting. Rising now to meet them past the hills, Kunming Airfield, foremost allied base in China, fans a dusty greeting from its desolate plate of packed brown dirt and gravel.

How important is this derelict outpost beneath the mountains? Only the chief strategic domino in China, short of Chiang Kai-shek's command at Chungking. Allied commanders know that if Kunming falls, Chungking goes next, then Chiang's lesser bases, and his government collapses. The ancient city with its alley-like streets holds few charms for Ned, for all its allure seven centuries past, as a hub of the Silk Road. Now a munitions manufactory, black-market conglomerate, and civic center for opium commerce and human trafficking, the place projects an air of delinquency and wartime neglect. It also boasts several decent cafes, many lesser ones, and a few dozen hotels, some leased by the U.S. for American servicemen whose quarters once catered to a resort trade of Chinese seeking balmy weather.

Whatever the city's cast, the place is a phenomenon of civilization compared to the base and its incongruous rock runway lying at 6200ft, an expanse leveled and refilled by coolies in human gangs, eternally shouldering into the pull of the huge rollers that keep the field at a semblance of smoothness, wielding their wonky carts, shovels, sifters, and picks. The panorama moves into the Commando's front as they labor down to roost, a sort of sensation of riding a gigantic pelican, with flaps and ailerons wide, wheels reaching low now for the floor.

"8209 to tower. Gear green, over." Ned increases the rpms on the props, and puts his mixture controls on auto-rich. "8209 to tower, reporting base leg. Over."

"Roger 8209, clear to land. Proceed to final."

Wheeling the plane from base maneuver in the crosswind, they move down final and touch earth at a roar. Ned signals thumb knuckle to forefinger—go light on the flaps—and works to reduce stalling. A minute later he's pulling into the ramp—so-named in the loosest sense brotherly love would permit—a metal-mesh haunt of dust devils, boulders as large as medicine balls, coolies with their baskets of rocks, and ground crews who move in to unload them. No need for a jeep. The flight line, base ops, and food are compactly clustered at hand. Ned and Brewster descend into the exotic ambience of Kunming, China, a primae-val outpost of asthma-invoking thin air, its surround of precipi-tous mountainsides broken, blackened, and scarred by aircraft that didn't climb fast enough to get above them, scattered patch-es of sunlight, mirey soaks of rainwater in puddles the size of fish ponds, and a living swarm of pilots and their planes.

"I'll sign us off," Ned half-salutes Brewster. "Meet you in the Egg Shack."

Coming through Operations, he notes a jokester's sign. *China is not for the timid.* So what clown ever thought it was?

Paperwork accomplished, now that blessed haven down the way, crammed with good fresh eggs. His mouth starts to water. For the timid or otherwise, China has a billion eggs, coffee strong enough to walk out across the picture-perfect foothills, soft rolls on your plate leaking their molten gold butter, a repast no king would find wanting.

In ninety minutes, the plane is unloaded and checked over, fuel calculated for the return, any "stray" drops drained. They've bought two beers which reign in lonesome brilliancy from inside the empty cargo bay, and which will be icy cold by the time they pass back over to Dinjan. They lift off with Brewster in left seat. Pilots put in a lot of copilot time in the Hump, getting others through or just swapping off, one seat going out, the other com-ing back.

They're climbing the eastern face of the ridge. Gotta hand it to Brewster, he's showing his stuff, the capable way he handles the plane, this man is gonna check out fast. They're into the Charlie

Route of the high Hump, getting into altitude and speed, into the usual goose-soup, as the summits of Everest's innumerable nippers and fry march in invisible disarray below. Ned's eyes rove the dials again, he leans back, reaches up and tweaks the radio frequency. Maybe Tokyo Rose had finished her discourse. Yep, there's Vera Lynn, singing We'll Meet Again.

"We'll meet again, I know we'll meet again
Don't know where, don't know when. Some sunny day . . . .
Keep smilin' thru, just like you always do,
till the blue skies drive the dark clouds far away."

Words to pull a man right in the old heart strings. To remind him how much he misses Neva Rae. Misses his mother. Wonders what his brothers are up to. Why, if old Rebecca came through the door tonight, he'd fall on her neck. Has Roxboro changed? Tomorrow-you-wait-and-see, he'll be visiting there, even if he figures to stride out on business, to wheel and deal and meet and greet. He'll come home to his wife and kids in civilian land, have himself a hot shower, make that a very hot shower lasting forty-five minutes minimum, then sit down someplace comfortable and dig into a few gallons of chocolate ice cream.

Must be a Vera Lynn party today. There she comes with The White Cliffs of Dover.

"There'll be blue birds over, the white cliffs of Dover, Tomorrow,
just you wait and see."

They're getting into some rough stuff, but the RPMs look good. He's seen plenty worse, plenty of times. Doesn't know if Brewster has, but he's sitting up tall in his mask, his hat pulled forward, handling the plane like a pro. A man who's doesn't lose his grip over a bit of a gust. The kinda guy he approves of.

He listens till Rose starts to fade. He could search around for her on another frequency, but he's been watching the wings, they're getting some build-up, gotta be sure the deicers are kicking in. You keep rechecking the whole panel anyway. Still can't see a blessed thing outside, and wouldn't you know, here comes the lightning. One, then another long jagged spear, right outside the cockpit window, so close they raise the hairs on his arms. He adjusts the rheostat so they don't get blinded. Brewster sits over there flying steady, a guy with ice in his veins. Immobile as ye cliffs of Dover.

He checks their coordinates. They're flying straight as a bullet. Gotta say it's a little rough today, bit of a crosswind, too, but they're making good progress.

By and by, he glances at his watch. Then the altimeter. 23,000ft.

Something wrong here, big time.

"Brewster?" His copilot sits gripped to the yoke, knuckle bones standing white, eyes boring out between his hat brim and oxygen like drill holes. Where're his gloves? Why's he still on autopilot?

"Brewster, prepare to make an instrument approach."

No response from left.

"Brewster." Is the guy deaf? Hardly moved a muscle, come to think, since that first lightning. It's going to take time to work down the slope. To get from altitude into Dinjan.

"Lieutenant Brewster, set up your descent."

Brewster hugs his seat, muffled in oxygen. They're going to overshoot their destination, and no fooling. Here comes another punch from the wind, this plane's getting pushed around, and they should be six thousand feet lower than they are, moving down in sure steady increments. Ned squelches the impulse to reach out, jerk away the man's face mask, and cuff him to attention. Instead he yanks away his own oxygen and raises his voice.

"Do you hear me, pilot?"

A spit of time, Brewster turns, peels his mask to the side, and reveals an ashen face.

"I—I don't know where I am," he chokes. "I'm lost."

Ned lurches forward, flips off the autopilot, reaches for the throttles and starts to pull.

"What do I do?" comes Brewster's dazed voice. "Where are we anyway?"

Twenty minutes later Ned sets the plane down at Dinjan airfield, from the right seat. No big deal. Just do the moves opposite like in a mirror, like when he was an instructor, and row it down hard. Only little problem would've been if they'd overflown the place at cruising altitude, and got themselves shot down on the way to Bhutan. Or maybe ended up out of fuel somewhere in a tea grove. Try explaining that to an unsympathetic commander.

A pilot who's never flown weather, they say, is like a sailor who's never been to sea.

He lets it go.

* * *

Ready to take command in China, Claire Chennault never expected to see the post jerked out from under him. But that July, 1945, in the worst blow life had dealt him, he found himself replaced by General Stratemeyer. Roosevelt's death may account in part. There wasn't a real consensus at the top either, but whatever happened brought a rough end to the career of the most-celebrated American in Asia. FDR had supported Chennault and air power, and his trust in Stilwell had been greatly shaken besides. In any event, Stratemeyer showed up with a blunt letter to General Albert Wedemeyer, from General Arnold: "General Chennault has been in China for a long time fighting a defensive air war with minimum resources. The resulting guerilla type of warfare must change to a modern striking offensive air power. I firmly believe that the quickest and most effective way to change air warfare in your theater is to change commanders. I would appreciate your concurrence in General Chennault's early withdrawal."[44]

The words overlook every plea Chennault ever made to push the air offensive in China. General Tunner, despite their argument over the appropriating of his Hump pilots, wrote with respect of this elder statesman and "old warrior," whose battle fatigue he notes with feeling.

Chennault greeted the news with a flash of his famous temper, then resignation. If Joe Stilwell was in any mood to laugh, he'd got the last one on Old Leather Face. His backbiting had paid off big. The only name he managed to smear worse was that of Chiang himself.

Accompanied by his Chinese wife Anna, Chennault spent a month visiting comrades and posts around the great country where he'd lived eight years of his life. He was greeted everywhere as a hero, a beloved man, one who would be irreparably missed. Out over the Hump, he flew to begin his journey home.

It was presently believed that the land he vacated was being mobilized for an all-out offensive, that new air strikes at Japan would be staged from within the continent. Men in the Hump could only assume a lot of their flying from now on would support the coming invasion.

General Stratemeyer's tune certainly softened for the work of these pilots.

"I am convinced," he writes General George, "that in years to come (this) command will receive an exalted place in the history of this war. The Hump is the most exacting and difficult air route in the world. The average pilot will admit that he would rather fly combat missions deep into Japanese territory than fly a heavily loaded transport over the Hump." [45]

Still cloaked in utmost secrecy was the test of the atomic bomb that same month, on July 16, at Los Alamos, New Mexico. Generals in CBI had no idea that the face of World War II, and of all wars to come, would be changed. Not William Tunner, for a certainty.

"(The) time set for the planned invasion of the islands was drawing near. We knew that in the assault on the fortified beaches, in the penetration of the islands which the Japanese could be depended upon to defend with fanatic determination, our forces were going to pay a terrible price. It was estimated that the invasion of Japan would cost a million casualties, with a high percentage of loss of life." [46]

The same broadcasts Americans heard at home came to servicemen over the Armed Forces networks. Bloodshed and carnage across the Pacific. The reconquest of the Philippines, with horrific slaughter there. Total completion of Ledo Road, pathetic in its stand against the overwhelming needs of Asia. Summit at Potsdam in July, with Truman, Churchill, and Stalin hammering out postwar control of Germany. Then ten days after Los Alamos, the U.S. issued an ultimatum to Japan: Surrender unconditionally or see their homeland destroyed.

For China, too, the hand of fate was writing on the Great Wall. By year's end the U.S. would be letting go hard-won bases and outposts across the vast old realm, and confronting tough questions as to how to honor their loyalty to Chiang Kai-shek. The Hump airlift, which President Roosevelt praised before his death as "an epic of the war . . . an amazing performance," would shutter down as abruptly as Chennault flew out of China. Generals Tunner, Stratemeyer, Wedemeyer and others, wouldn't be needed for air missions carried out from the mainland. Events were poised to sweep a lot of players off the board.

Only not the U.S. Hump pilots. Not yet.

China accelerated demands for help and materiel. Two journalists on the scene, Theodore White and Annalee Jacoby, make no attempt to gloss the truth. "Strain as they might, die as gallantly as they did, the airmen of the Hump could never meet the insatiable voracity of the beleaguered garrisons beyond the mountains." [47] Life went on in a grueling clockwork of danger-fraught flying and proximity to an enemy who showed no disposition to surrender. Every crossing hurled pilots back into the turbulent darkness that carried death on its wings. There are no atheists in foxholes, men say. Nor probably over the murderous maw of the Hump.

"I'm a believer, and well, you have those moments," Ned admits. "Not that I could leave the controls, go back to the cargo, and throw myself on my knees."

By July, tonnage flown across the Himalayas again exceeded all records. Every two to three minutes another plane crossed the range. Like soldiers inching their way over battle-torn terrain, the pilots kept coming, wave on wave. The men who flew, the crews who kept them flying, a hard-driving general who made airlift a science and worked the details down the grid, all gave everything they had. General Tunner tasted his victory. Many had died but his vision was paying off. He couldn't banish the dangers, but he'd managed to cut waste and raise the bar on safety with meticulous flight procedures, and by pushing his command to the limits.

With Army Air Forces Day on the approach on August 1, he decided on a new mission. General Arnold had wanted to observe the day with some parades and galas here and there. Tunner, however, was decidedly not in a gala mood. Estimating the terrible cost of life to come in the invasion of Japan's beachhead, he saw little to celebrate. Instead he decided to mark the occasion, all 24 hours of it, with a push to fly more supplies than ever before across the Hump.

Some celebration, to spend a holiday working at maximum effort. Luckily for Tunner, he was betting on the pride of healthy men and their American competitive edge. Even commanding officers would fly. He himself flew three round trips at full throttle. Granted, he was never in the pilot's seat. Supplies and materiel were readied to load fast. Every base was mobilized to have a plane fresh for takeoff within moments of a landing. Betting pools formed, airfields got prorated statistically for the aircraft they could handle and move, odds made fair all round. A lot of fellows dug into their wallets. Money was going to change hands on this one.

No one slept much on the Big Day, that's what they called it.

But could pilots already flying so hard really excel themselves in a single twenty-four hours?

"They wanted to break all the records, and we did it," Ned affirms.

Let's say, too, that for once the abominable rains held back as if to strew a benediction over the effort. Yes indeed, when pigs fly. Ned reported to base ops around eleven at night, in the usual dark thick downpour that had his trouser legs soaked to the knees. He was greeted by a couple of Red Cross ladies presiding at a table spread beneath the dripping thatched roof of the tower porch, handing out coffee and doughnuts. It would have taken an industrial sump pump just to sweep the walk.

"They made a gesture, but I didn't usually drink coffee before flying." Few did. Not in planes lacking toilet facilities. The operation would extend by military time from 0001 to 2400, that is, a minute past midnight to the following midnight. Ned's midnight takeoff for Kunming put him among the first pilots off the ground.

"I went out with my copilot, we cranked up, got into our masks, and had the radio on listening for the tower. Then we waited on the line to go over Hump. Runway lights were a blur in the rain. We could just barely see them when we took off, and we were into fog pretty fast. I made good time. Getting back near base early next morning, I was able to get above the clouds for a while. I remember the sunrise across the mountains at my back."

General Tunner chalks up the day with near-boyish elation. "The statisticians had worn out their pencils figuring up the totals. (The) India-China Division had made history on Army Air Forces Day—1,118 round trips over the Hump, with a payload of 5,327 tons. This averaged out to just a little over two trips per available airplane. One plane crossed the Hump every minute and twelve seconds. Four times a minute a ton of materiel was landed in China." [48]

Were safety standards ignored? No. Pilots were mobilized

who might otherwise have had a day off but they didn't fly twenty-four hours, nor could have. Ned considers twelve the outside edge a pilot can push before his faculties grow drained, but even that's too much. The most he ever flew at a stretch was sixteen, under exceptional conditions, "and I was pretty wrung out." Men flew a round trip each. Tunner's object wasn't to get them killed. A lot of the day's effectiveness came of having procedures streamlined to a fare-thee-well. Camaraderie and team spirit were felt, and everyone greeted the best statistic with enthusiasm —zero crashes or "accidents" that day on the Hump.

Life returned to business as usual till August 6. Some of the fellows were swapping greetings in the basha, the radio playing, when news broke over the armed forces network.

The U.S. had dropped an atomic bomb on the city of Hiroshima, Japan.

Ned recalls the dead silence that fell among them.

"We had no real idea what it was. We couldn't imagine the amount of destruction. It was hard to take in. It was the first time we thought we might foresee an end to the war, and that men would be coming home. It crossed our minds to wonder how many more flights we'd be making across the mountains. We didn't know we'd be going to China."

No pictures had hit the newsstands to show the flattened, devastated expanse where a city had been and people lived. Colonel Paul Tibbets, who captained the *Enola Gay*, would say he never regretted the deadly mission he commanded; too many Americans had been killed, tortured and mistreated by the enemy. Subsequent intelligence suggests Japan was hovering on surrender, only delaying to argue over high-level consensus and iron out details regarding the fate of their emperor. Still, no word of surrender came.

Whatever Japan's internal dilemma, and in mounting apprehension of the fearsome loss of life invading Japanese shores, on August 9, the U.S. bombed the city of Nagasaki.

On August 14, President Truman received the unconditional surrender he'd demanded. The peace became official two and a half weeks later, on V-J Day, September 2, 1945, when Japan signed the formal armistice in Tokyo Bay, on the deck of the U.S. Battleship Missouri.

* * *

Across the measureless sweep of Asia in the dwindling summer of 1945, no one could look in any direction on the compass, and see peace in the valley.

Hostilities erupted into full-blown civil war in China. Divested of the Japanese foe, Chiang Kai-shek had cause now to expect the Soviet Union to throw support behind Mao's Communists. The Allies had won the war. His own hopes for victory appeared far from sure.

His ragged troops held on between the frying pan and the fire. In their midst hunkered the menacing presence of 2 million surrendered Japanese soldiers, men who lived hale and well-fed thanks to their long ruthless pillage of the people around them. The enormity of packing them home would take months to resolve. Three-hundred thousand quartered in Shanghai alone. With General MacArthur's army tied down occupying Japan, no military force on the mainland could corral the conquered, who in consequence saw no cause to alter their impudent, sneering attitude toward the victors. For a time, no one was sure if the enemy troops even would honor the defeat of their emperor, so firm was their hold inside China. With various contingents delaying surrender for assorted weeks after Armistice, they continued to live, as they say, very high on the hog among their new lords.

General Tunner put his whipsaw brain to cracking again, and proposed a long-range solution he called his Orient Project. Conceived to link up vast regions of Asia - a territory divided by innumerable languages, peoples, and customs, minimal infrastructure, and nonexistent communication—in this massive good-will undertaking, airports would be built and community fostered, via the air.

He had his pilots and air fleet ready, and the native manpower to build airfields. He'd replace older aircraft with spanking-fine C-54s, and lay out modern runways to greet them. The way he reasoned, men gladly would bring their families out to the Orient, and stay on to accrue the flying time that would insure lucrative tickets into commercial aviation when they returned to the States. Operational officers, men from Major Langford all the way to generals with a star or two, could enhance their careers and managerial skills by administering interesting new responsibilities. It was an ambitious, beautiful idea, and even commanders on the scene expressed interest in thus fostering commerce and understanding across the strife-torn continent.

What Tunner didn't figure on, and maybe should have, was the outcry back in the States. America was sick of war. Families who wanted their husbands, fathers, brothers, sons and sweethearts home again, besieged congressional offices. Senators and congressmen made promises right and left to get the boys out of Asia and the Pacific, and home for the holidays. They rang the phones off the hooks of military commanders to get the fellows headed stateside. No one even wanted to hear about an Orient Project.

As Tunner's soaring dream got shot from the sky, the gritty nightmare of the Hump droned on. Men continued to fly the Himalayas after Hiroshima—records indicate at least into November, well past the time when the Airlift officially shut down. Who can wonder that a renewed spirit of gallows satire took hold in the place? Had Americans trounced the enemy only to crash and die on the Aluminum Trail now as victors?

Then came a small aftershock. Those who'd been longest in the theater rotated back to the States, but for pilots remaining in the Tenth Air Force's Troop Carrier Squadrons came a new mission.

Orders for deployment farther east, to fly for Chiang Kai-shek.

Ned didn't complain. "It was only fair that those who had put in more time should go home, and those with less should stay." Ranged inside the dark bellies of their squadron planes, the men flew by successive detachments into China across the lower Hump, Ned being among the first out in September. The pilot was a personal friend, a lanky good-humored flight officer of twenty-six, Si Singram, from Kingsport, Tennessee. Along the familiar skyway there was a lot of comradely back-and-forth between cabin and cockpit. Beside Ned on the canvas sat Jack Mayer, in top spirits, a first pilot now, his past difficulties forgotten. Crossing the roof of the world, it was a first flyover as passengers for these seasoned young men who'd fought the range for life or death as pilots.

What kind of work waited in China? For a start, thirty thousand Chinese troops had to be airlifted into Shanghai, to accept the defeat of Japanese servicemen and begin to administer the city. Ned, Jack, Si Singram, and the others would be doing a lot of flying into Shanghai, where aside from a small commercial airport, they met with a pitiable show—and no fault of the Chinese, who suddenly had to muster and take charge of an enemy who'd lived eight years at the top of the hill, who walked about the city well-fed and impudent, and saw no call for subservience to the smelly, barefoot, louse-ridden beggars Chiang Kai-shek ordered in.

So the story went across Asia, from Peking to Hanoi, a starved and demoralized winner trying to bring order to the comfortable, unrepentant vanquished. Without a doubt, one of the greatest blessings the Nationalists had going for them was the U.S. pilots who deployed to help give their leader and Generalissimo another chance.

Chiang Kai-shek desperately needed one. A new and massive air operation took form overnight, this time within the frontiers of China. General Wedemeyer, under orders from Washington, drew up the plans whereby U.S. pilots would fly Nationalist troops by the hundreds of thousands from the south up into Manchuria and northern provinces, to fight the Communists.

Ned's first posting at Chihkiang (now Zhijiang) happened to have been the staging point that May, from which the Japanese massed a harsh offensive that roared brutally along up to the time of their country's surrender. Arriving on this scene of bloodshed, Major Langford and his squadron hardly found a spot in frame to accommodate them. What met their eyes instead were enough Japanese soldiers strolling abroad to dwarf the lonely outpost of Americans, a scene common to all the bases, as Ned remembers.

"We often saw the Japanese at the fields when we landed. The war was just over and they hadn't moved them out. The officers used to walk around with these big swords swinging at their sides." How to dispatch this erstwhile foe, the same they'd watched out for, air and ground, through the Hump war? One far-fetched possibility was that China might employ these idling troops to help fight the Communists. But in the end, it seemed too risky to reinvest power in so recent an enemy, however much in fighting trim they appeared.

Where to put a squadron of Americans pilots, practically speaking? With the Japanese living in comfort in their own billets, why not do them better with some spiffy BOQs and a mess, maybe a clubroom or tennis court? Afraid not, not in Chihkiang. A few dozen canvas tents sprouted across a field near a ravine. Pilots who'd never used helmets now bathed and brushed their teeth from them. Mushy meals, cloudy water, the old jaundice-inducing Atabrine tablets, starred in the chow line. On the positive side, sweltering beds and dripping cockpits were things of the past. Flung along the other edge of the ravine lay the one trace of modernity in sight, a rutted dirt runway ready and waiting for big U.S. planes.

General Tunner left his former quarters in Karachi, crossed the Hump, and relocated to his forward position in Kunming. His pilots fanned into assorted holding patterns as bases were readied for them. Wherever they posted, they had their work cut out. Some of the sorriest stories to emerge from China involved the transport of Chiang Kai-shek's troops. On the ground at Chanyi, Tunner personally witnessed an incident where a panicked unit of soldiers rushed en masse to the rear of a C-46, speeding into takeoff. Hoping to jump out and escape, they unbalanced the tail and caused the aircraft to flip over and crash, killing everyone on board.

Every pilot in Asia was on the lookout for the dragon curse. Superstitious coolies wishing to rid themselves of assorted demons and dragons on their tail, would hurl themselves across the runway in the path of a takeoff, the object being to clear the propellor on the near wing, crouch under the belly of the roaring aircraft, let the rpms of the prop cut off the dragon behind them, and grind it to a sausage. This didn't always work, and coolies got pulverized in the whirling blades, incidents that unnerved and sickened pilots, who crashed planes as they skidded off runways trying not to murder some poor fanatic at takeoff speed. General Tunner sent an order from headquarters: Maintain runway momentum dragons or not, but the "accidents" continued, and in one of the strangest missives of his career, he requested to classify "dragon deaths" as something other than accidents so his bases wouldn't have to keep reporting them.

Some Chinese soldiers were hardly but boys, rounded up from villages and paddies, and impressed into military service. They and many of the men had never flown, and were terrified. Pilots learned to close and lock their cockpit doors, lest troops blame them for rough rides in turbulence, and turn to attack. Soldiers might storm to the front of the plane, or run to the rear, or thrust their bayonets at walls, crew, at whomever or whatever they could target. In calmer moments they built campfires on cold fuselage floors, brewed tea and heated rice. They played catch with hand grenades and shot off rifles, riddling planes with bullet holes. Fights and brawls broke out. Cargo doors had been opened at cruising altitude, from which were pushed out to their deaths a comrade or two, something pilots were powerless to prevent, and typically didn't discover till they landed with fewer passengers than they'd taken on board.

On a flight from Sian to Kunming, also recounted in General Tunner's memoirs, rowdy soldiers nearly battered their way into the cockpit before the pilots fastened on their masks, pulled the plane up, and their attackers dropped off from oxygen deprivation. When they landed, "they were all still suffering from anoxia and debarked like sleepy little lambs." [49]

A nauseating stench blanketed the ramps. Peasant soldiery defecated and urinated on aircraft floors, and vomited up meals when air sickness hit. Interpreters were dispatched pre-flight to instruct them on airplane etiquette; more effective may have been the addition of a few well-placed buckets in the cabin. Word went out that weapons would be locked in the belly bay of the plane before troops boarded. Pilots found themselves disarming unruly troops, fifty to sixty men for a C-46. Confronting these mobs, Ned himself refused to fly until compliance was achieved, and often personally stashed their firearms below his trapdoor in the cockpit.

He made no complaint of the smells. Bathing was hard enough for Americans. A matter that would stay with him—neglected in most accounts—was the latent realization he'd been flying "comfort women" along with troops, an aspect of warfare that had not in those days received scrutiny as a human rights outrage. How many women and girls were kidnaped into the abhorrent slavery? Shrouded in mystery to the world, what happened to women in China at the hands of the Japanese, as to an extent also by their own Chinese brethren, must pass down as one of the cruelest abuses in human history. Looking back, Ned says,

"To tell you the truth, I couldn't distinguish the women from the soldiers, with everyone so bundled up, loaded down with gear, with clothes and pots and blankets. They had to carry everything with them. I felt bad for them but I knew I had to fly. More than anything in those days, I felt terrible for the children - starved, in rags. It broke your heart."

Now if there were a thing with which Ned Thomas happened to be gifted, it was a warbling whistle that could burble across a field like a cheery brook or a happy bird, and no one else on earth could duplicate it. So it sometimes came about that autumn, the warble would break forth along some narrow street in Peking or Hankow, Chihkiang, Shanghai, Kweilin or Liuchow—for tried-and-tested Hump pilots flew across thousands of miles of China's lower back. And if you followed the whistle, there would be a cluster of children shyly smiling at a tall American who could make a Hershey bar go all round, and break into a nutty little jig before their widening eyes, "my old soft shoe," Ned called it, Carolina on Asia soil, the tapping and jiving of a Roxboro kid who'd acquired it on the fields and alleys of his childhood with friends whose color may have been his, or maybe not, because they were all just having fun. Then for a little bit of time, a few dozen small brown faces split into smiles and had fun, too, and happiness settled over a desolate byway of the war like a wreath.

In October, the men's APO changed again. Major Langford's First Squadron flew their '46s from Chihkiang to Hankow—present-day Wuhan—which lies on the long, shining Yangtze, a hub in days of yore when clipper ships had sailed there to take on cargos of tea. In 1938, the city fell to Japan; in December, 1944, additional portions were destroyed in U.S. firebombing raids by the Fourteenth Army Air Force. A bit of the old graciousness survived in the British concession area, and after the fall of Nanking, Hankow became China's wartime capital. Of course, none of this ambiance got extended to Americans. The runway was dirt and grass and plenty rough, save for a short paved stretch. Pilots took up quarters in a bombed-out hangar absent a proper roof, and by mid-month, sleeping conditions were frigid.

"I went to bed with everything I had piled on top of me," Ned remembers. At least, the food improved—plentiful eggs, vats of butter in the mess kitchen, and fresh-baked bread. They were still in a war zone, as if anybody needed reminding one memorable night.

"A Japanese ammunition storage facility was there on Hankow base. Someone went out, we were never sure who, armed one of the bombs, and the whole dump blew up. Aerial bombs have that sort of propeller on the front. We saw tree trunks, all kinds of objects coming down through what remained of our roof. An explosion like that, it gives you a shock, so what do I do?" Ned laughs. "I dived under my canvas cot."

It was at Hankow, on October 15, that Ned's promotion to First Lieu-tenant came through the wire, changing the single bar on his shoulder from brass to silver. A friend passed along a bar for him to wear. He was pleased to see Flight Officer Si Singram, who'd flown them out of the Hump, get a promotion to 2nd lieutenant, and to him Ned passed his own brass bar, as men often did, handing along wings and insignia to companions in service, as their own status changed. His first wings, on the reverse of which he'd scratched his name with Neva Rae's in Malden, had been given him by an instructor at Moody.

Men still wore the rank of flight officer proudly. Despite this, Ned had reservations about the designation. After all, they performed the same work as commissioned pilots. Still, the military functions on respect for discipline and protocol, as he knew. "Your peacetime training manifests itself when you get in wartime conditions. You've got to have standard operating procedures under combat-prevalent conditions, and certainly in a squadron. Nothing is left to chance." Because when things fall apart around you, knowing what to do saves lives. A clear divide is drawn between enlisted ranks and officers—discipline can't be maintained otherwise—but Ned's attitude included real respect for the efforts of all.

"In the Hump, the barrier softened some, though we ate in separate messes and slept in separate quarters. Our enlisted ground crews were as important as any officers or pilots there. Without them we couldn't have flown, or flown safely."

General Tunner, a stickler, may inadvertently have contributed to one important breakdown in procedures—sloppy record-keeping—by decreeing that pilots didn't head home till their tours ended, regardless of hours flown Where hours didn't matter, clerks grew lax, and men often ceased to verify paperwork, unlike those who came earlier who may have kept their own logs, serial numbers of aircraft, and names of crew, knowing when they achieved the magic 650, they were headed stateside to collect a Distinguished Flying Cross. Flight records called Form 5s were so incomplete in the Hump, that it might be asked if the information vaporized in the gut of the mountains. That the military credited Ned with 350 hours of Hump combat flying, in excess of 100 hours a month, for all the odd gaps in his Form 5s, and where 65 hours were considered the safe maximum, raises unanswered questions. Before he returned, he and records staff reckoned up some 300 hours he'd spent as a pilot, trading off and logging another 400 as a copilot, and these hours had to take place somewhere. Where, the papers don't say. He knew he flew a lot of missions. Like others, he simply got up and persevered.

"It was my job. You didn't spend your life losing sleep over something you had to do."

Being young and indestructible, pilots in the First Squadron found one boisterous way to break up the routine—and any job may have seemed routine to men who'd faced and overcome the Hump. Ned describes how they'd radio each other from their cockpits, marry up their wings, and fly in formation as they homed into Hankow, a fleet of '46s in a wedge across the sky, spilling down to base and the airfield. Major Langford either approved, or stroked his clipped mustache and let it be.

By now Ned had circled more planes down into Peking than

he could count, finding himself in growing wonder of a great warren of aged buildings sprawled below his view. As his plane dropped in increments, he'd see the ancient bricks, the winding streets, a grand and somber center with ruined gardens and what could have been . . . a moat?

One day after flying in another contingent of Chiang's troops —who debarked thence for northwesterly hinterlands, dark and strange—and while his plane got serviced, he made up his mind to slip away and investigate. Hitching along in a bouncing truck carrying some noisy GIs, he attained a stretch of road guarded at either side by outlandish legions of statues bigger than lifesize, dragons of stone, lions with claws unsheathed, phoenix-like unnamed assorted beasts.

At the walls, he alighted alone. Best to stride along purposefully, as if he knew where he was going. Which he didn't. It just was prudent for a man to look purposeful in a war. To look like someone who'd got his promotion and was the pilot in charge, and who, by golly, that plane wouldn't be going anyplace without, till he got back. Having a bit of time to wander and check around, he directed his steps in a centered manner, not that he had a clue what center was either.

More confounded did he grow with every step, at the archaic scene that met his eyes.

Peking was no metropolis. Shanghai was far and away bigger. Moving steadily into a place of rose colored walls, of walls looming within more walls, he glimpsed the moat afar, for so it proved, low lying and reflecting the light of the sky. Encountering fewer and fewer people, by and by, his steady pace thrust him out upon a great blanket of space.

Dry leaves gusted across the weathered and deserted pavers, the only sound in the cool brisk air, while a lone American turned slowly on his heel and gazed in wonder.

Before him beckoned a vista of stairs rising to a great sort of hall sheltered over by two enormous pagoda roofs, one above the other, fancifully adorned. Flung out at every angle stood a vast surround of ancient constructions, of rosy weather-baked bricks, of porches adorned with fantastical trim work, a maze of corridors dividing one structure from another, tiled roofs with bizarre beasts guarding their slopes like grimacing, coiling sentinels.

This was no fortress he was in. Nor even a town, but an honest-to-goodness palace. Despite the thick walls guarding the whole, a spacious feeling opened, wrapped in eerie quiet, like a kingdom set apart, like some hidden enchantment, where time had stopped.

"I was in the Forbidden City, and I didn't know it. I can't say if Americans even called it that in those days. I looked around, it's impossible to describe. All I can compare it to is some scene from The Last Emperor. I'd heard of the Boxer rebellion and the opium wars. Never anything like this. And there was no one to tell me what I saw."

The old Forbidden City, imperial home of the Ming Dynasty. Acre on acre of heart-stopping silent beauty in the heart of what is today bustling modern Beijing, China, a World Tourist Site. Where for an autumn hour in 1945, an American pilot from North Carolina had the place in solemn quietude, just about to himself.

CHAPTER 7

"IF THE PLANES WERE READY, WE FLEW"

PILOTS WHO DEPLOYED FROM A WORLD WAR IN THE HUMP, INTO a raging civil conflict on the Asia mainland, could hardly have been faulted if they thought there wasn't much farther for China to fall in her race to the bottom. By now Chiang Kai-shek had acquired his reputation in the west as a double-dealer who'd play anyone false. "General Cash-My-Check." How much money had "Uncle Chump From Over the Hump" shelled into his war?

In one respect, Chiang hadn't changed. His heart burned for a republic. Standing before the legislative body of the Kuomintang, he presented his resolutions to end single-party rule in China, to protect divergent political voices, and institute free elections in which every adult over 21 would participate. To his crushed people, he promised a constitution and such democratic rights as an 8-hour work day. [50] Three years later, in 1948, he assumed the presidency of the Republic of China, an office he held about a year and a half, all told.

For him or against him, no one could deny he'd spent most of his life waging war, his early clashes with the old feudal warlords, then with Japan, then Communism, all in a failed effort to unite his country. Instead of the mainland republic of his dreams, he eventually would build his vision in prosperous Formosa or modern Taiwan. As for Mao Tse-tung, it took decades to uncover the depravity and corruption of this, China's so-called peasant dictator.

Whatever reforms were talked about, life was dirt-cheap in post-World War II China. Ned looked out on scenes that seemed to regress to prehistory—the desolation of a war-torn, exhausted land where the knell of death pervaded at every turn. In later thoughtful moments he tried to assess where he'd stood in these experiences as a pilot and officer entering his twenties.

"I found myself in a culture thousands of years old, and I wish I'd been smart enough to benefit and capitalize on the opportunity. The Chinese homeland was being devastated. We saw the poverty and suffering. But there was so much we didn't have time to learn about. We were in the air. There was no way to get to know the people.

"We used to take a jeep into Hankow for a meal. The city had been badly bombed. We encountered people existing in shacks, the terrible destruction. We'd go in a restaurant, find a bare room with one light bulb hanging down or a lantern. The only water we could drink was on base, hanging in an olive-drab Lister bag with chlorine to clean it up—the things kinda sweated, and we'd fill our canteens there. We carried a .45 shoved under an arm, and we towered over the Chinese like aliens. I'd see these elderly women with canes, hobbling along the streets on misshapen stubs from foot-wrapping. A shame to cripple little girls that way. You were careful about the food because you could really get sick. Taking off in a plane, I'd see farmers heading out to the fields with their honey buckets. It was a way of life.

"There you are in the cockpit, your engines flinging out sand from Chinese airfields, and nothing deteriorates machinery like sand. What you hope is to make another safe trip to the front where Chiang's massing his armies, and you've got half a hundred troops hunkered back there in the cabin, and they could break into a riot and turn the plane over. I tried to do the job, and

I did it safely, so to that extent I guess you could call it a success."

No imagined glorification of war comes through his words. But no antipathy for the Chinese either, even an intrigued interest for their culture and ways that he had no means then to satisfy or study. The co-authors of *Thunder out of China*, Theodore White and Annalee Jacoby, having observed American military men in the theater, paint with a broad brush in naming contempt as the abiding sentiment of U.S. servicemen for the Chinese people: "They saw the squalor, filth, and ignorance of the Chinese peasant and peasant soldier; the sight inspired them not with compassion or pity but with loathing and revulsion." [51]

Not for Ned, not in his growing awareness of pity and sorrow.

* * *

Airlifts would have their likenesses, yet each presents unique challenges. The Hump was the first, and without a doubt the worst. Berlin, another example, involved a lot of instrument flying, and pilots had to know how to do it, but the designated route was shorter, over low terrain and on a radio course. So it went, but no matter where they flew from now on, American Military Air Transport, or MATS, would be, in General Tunner's words, "the bellwether of international trouble. . . When trouble breaks out, it must of necessity be MATS which first responds to the emergency calls. MATS must have the planes ready. MATS trains 365 days a year; it is always ready for D-Day." [52]

Tunner's vision—airlift a science and a specialty—became the reality, thanks in part to his fierce focus, if equally to the tough, persevering pilots he commanded. Gone were the days when air transport was left to luck or to whoever happened to be on the scene. It took expertise to tend the minute concerns that safety, success, and speed required. The Hump brought this home. The Berlin Airlift was on the horizon, and there Tunner would take his CBI experience to achieve another triumph in aviation. In a few years he was at the helm for the Korea mission.

Was the main achievement of the Hump then, to anticipate lifesaving flights into Berlin and elsewhere? Not really, and here was the paradox. While synergistic with America's war in the Pacific, at the same time the Airlift backed a struggle in China that failed, and a leader who could not surmount the odds he faced, despite champions like Claire Chennault and all the crews, living and dead, who'd braved the murderous Himalayas. The Hump—undersupplied, underserviced, crippled by infighting at the top—succeeded in convincing Roosevelt and commanders like Marshall, Arnold, Mountbatten and others, that no earthly track anywhere could achieve what air power could do. Time proved them right.

When England's Winston Churchill chimed in, he was in full throat, if ever he was.

"(The) U.S.A. has been increasingly engaged in establishing an air route to China capable of carrying immense supplies, and by astounding efforts and at a vast cost. . . over the terrible Himalayas or Hump as it is called, I would not say how many times as much as the Burma Road ever carried in its palmiest days. This incredible feat over ground where engine failure means certain death to the pilot, has been performed by a grand effort which the U.S.A. has made in their passionate desire to aid in the resistence of China. No more prodigious example of strength, science, and organization in this class of work has ever been seen or dreamt of." [53]

Over the Top of the World in a "forgotten war," modern strategic airlift was born. Was it worth it? Chiang Kai-shek couldn't secure a victory over Japan, but what others knew, he knew also. He was no fool, and he could always aim to hold on till greater powers brought the enemy down. He did hold on, and America gained. Symbiosis. Though perhaps U.S. troops who'd served in the Pacific best appreciated that in "accomplishing the impossible," the Hump kept the Generalissimo on the march and spared American blood. General Tunner drummed the point home: Thanks to the Hump, Chiang mounted a sufficient resistance to force Japan to keep a formidable force in fighting trim in China, and every enemy soldier there, was one less in the Pacific. He writes.

"We flew that airlift over the highest mountains in the world, in good weather or bad, over large areas of territory inhabited by the enemy and by savage tribes, even head-hunters, and with a confusing variety of planes. Through this airlift, and it alone, we kept sixty thousand American soldiers and nineteen Chinese armies sufficiently well supplied to tie down the Japanese. All the Pacific campaigns, tough enough as they were, would have been that much more costly in American lives. We flew almost a million tons of cargo over the Hump, including food, ammunition, and gasoline, mules and steam rollers, and four Chinese armies.

"Remember that once the airlift got under way, every drop of fuel, every weapon, and every round of ammunition, and 100 percent of such diverse supplies as carbon paper and C-rations . . . was flown in by airlift. Never in the history of transportation had any community been supplied such a large proportion of its needs by air. Yet this was achieved in the Himalayan Airlift, undertaken with no previous experience and under the most difficult conditions. Begun when air transportation was in its very infancy, carried on with steady increase in spite of the enemy and formidable weather conditions, and over the most menacing terrain, all this half a world away from home, the Hump Airlift proved forever the efficacy of air transportation." [54]

* * *

For pilots in China, Ned's old words rang true. "If the planes were ready, we flew them." Flying hours, human limits, aging aircraft be hanged. The objective was a last all-out effort to rearrange the far-flung Asia chessboard. Everyone knew it was airplanes that moved the pawns in the game, pilots who moved the planes, and as it happened, both were racing time.

Why? History again. Having rejected Chiang's overtures to negotiate, Mao Tse-tung dug into Manchuria and mustered his grab for power. If his gamble paid off—that the United States was too war-weary to challenge him—it would be checkmate. He was betting, too, that sooner or later the Soviets would support him ideologically, if not cover his back militarily. The business played out in the oddly familiar way where Mao, just like Chiang before him, would bide his time. Stalin wasn't about to confront the sole atomic power on earth, but this was bound to change when the USSR had a bomb of their own, for it was manifest that even decimated by war, Russia had every intention of becoming the world's next atomic stronghold.

By the terms of the Armistice, Japanese troops in China could surrender only to Chiang Kai-shek's armies. The stipulation created another near-unimaginable job, to get hundreds of thousands of Nationalist soldiers moved around to where they needed to be, and do it fast. Even with Chinese and American planes flying at capacity day and night, transporting so many men around so immense a map presented a whole new challenge for aviation.

Could overland transport have been marshaled for the task? Not realistically. Roads, vehicles, fuel, infrastructure to move huge armies rapidly across the sweep of China, just weren't there, a dearth that was one of the Generalissimo's greatest nightmares. To march armies where they needed to go would have involved months, and they had to be shifted up and down too fast.

General Wedemeyer, like his government back in Washington, must have felt hemmed in by a wall of fiery dragons. Uncle Sam had drawn the last line. The U.S. would not embark on a struggle that pertained to China's internal affairs. Yet to walk away and forget Chiang's years of service and America's debt to him, to abandon him to the Communists, seemed unthinkable. Wedemeyer himself labored under no illusions about Mao, a totalitarian who endangered not only China, but vital future interests he assessed for his own country.

As the clock ticked louder, order after order did Wedemeyer sign to move Nationalist troops along by every means at his command. America hacked back aid and assistance, he bent the rules, worked the men, and interpreted orders as free-handedly as he dared. Tunner's dream to thread Asia together by airplanes and good will, whether practical or quixotic and illusory, was dead in the water anyway. Small wonder pilots were flying hard.

Had the US wished to take a page from the Asian mind, an ambiguous approach might have been considered. Perhaps a waiting game, long enough anyway to figure out the next move. What happened instead came in a discordant rush. Roosevelt's envoy and Ambassador (former Secretary of War) General Patrick Hurley abruptly resigned over the disorganization and insubordination he encountered in the nation's Department of State. Downplaying the Communist threat somewhat, he had not been Chiang Kai-sheks' greatest ally, but he also was honest in his assessment that the U.S. had shamefully undercut the Generalissimo at Yalta. He fully supported the Nationalist government as legitimate, and had every expectation that the Communists ought to come around to the same. He was heartily fed up as well, with crisscrossing the globe from Washington to London to Moscow, trying to deal with implacable foreign service officers who viewed China as a doomed cause, and a purely American problem.

Brought up short by Hurley's departure, if not by sobering reminders of China's strategic importance, President Truman called General George C. Marshall out of retirement and asked him to go to China in December, to work for concord between the Nationalists and Communists, and help sort out the mess. Truman saw America edging near a whole new quagmire in Asia, yet whatever happened, he was determined to keep the U.S. from falling in and drowning there.

Marshall's scheduled arrival a couple of months distant was in no way likely to alter the lives of U.S. pilots in Asia. That die had been cast.   They just had more work to get done on their way out.   By autumn of 1945, flyers moving Nationalist troops simultaneously were transporting American servicemen into Shanghai to begin their journeys home.

What about starting them home by flying them back across the Hump to India? Not on your life. The route was too risky, too perilous. Another irony of that dark, forgotten war.

The old harbor of Shanghai told the new story. American ships lay ready to ply China's waters, filling every berth, getting set to sail the men stateside.

How did General Wedemeyer feel as he razed down bases and dismantled and junked military equipment and sturdy, airworthy planes for which his own country, too, had paid so dear? Sick probably. But he had no choice.

* * *

November 25, 1945. Sunday.

"Who'd have thought, Ned, you and me, we'd be the ones to close down China? Practically all by ourselves. One last pit stop here in Kunming, and we're home and dry. Why, we just about personally handed Hankow back to the native help yesterday."

Ned shrugged. "The handing got done way up over our heads."

Crossing the snow-dusted Kunming Airfield, Ned Thomas and Jack Mayer adjusted their garrison caps, their brogans merged in stride. Broad-shouldered Ned stood a shade taller, his whip-thin companion ruddier about the face, on both lay a roughhewn shadow of gauntness, what with monotone K-rations eaten on the fly, and the want of sleep that pooled in Ned's brown and Jack's steady green gaze. Late last night the two had deadheaded in with a final assignment. Pick up a C-46, fly it back east to Shanghai, and turn it over to the Chinese military.

The big plane stood among a smattering of smaller aircraft flung around the field beneath a bleakly overcast China sky, which today was huffing a lot of cold blustery wind.

"Well, we gotta watch our backs extra from here on out, 'cause we're jinxed." Jack zipped his brown leather flying jacket. "There's no worse luck than to be the last guy out of anyplace. That's when the cards have it in for you. Now I didn't see any other Americans flying out of Hankow when we left, did you? So picture this. Family of some poor sap gathered back in Podunk, their hearts fluttering for the sound of his knock at the door. Then who shows up? The telegraph boy. With a wire from Uncle Sam. *Sorry-folks-Stop-Oscar-got-his-today-Stop.* Yep, old Oscar bought the farm. Just when he was flying the last plane out."

"You're assuming ours was the last plane," Ned suggested

"You see any other guys in our squadron milling around? Si cruised that gooney bird up from Chihkiang yesterday, didn't even climb down. Just heaved the door open, said come right on in, boys, my home is yours, and took off. Langford beat it outta there around noon with his files stuffed in some rocket boxes. And get a load of that holiday shindig he's throwing for us. At the Kavkaz Café. Toniest nightspot in Shanghai. He's got Russian cocktails and filet mignon on the menu, with dinner music and a floorshow, all to honor our squadron before we ship out. Too bad we won't be alive to enjoy it."

Ned cast an assessing eye across Kunming field. For the times past number he'd flown into its hubbub, a feeling of eerie deser-tion enwrapped the place. The noisiest base on the map, now nothing cut through the strange, unaccustomed quiet. He took in a clump of P-40s, a derelict '47 with fading camouflage at the revetement, farther off in some weeds, a Liberator with sand blown around the wheels, looking way overtime for service, insignia flaking off, hunkered down peeled and sad. November was closing down cold as the grim reaper's toenails, they'd been moving planes and men nonstop, and the crop had thinned.

"Okay, so we're the last," Ned conceded. "So what? Let's get in and crank up."

"Good thing, I got my lucky nickel in my pocket. Maybe our

luck hasn't run out." Jack stopped mid-stride. "Pilot or copilot, match you for it."

"Flip for it," Ned answered. "Pull out your nickel."

"You'll wish we'd done matches. It being my lucky one."

Jack tossed a glint of silver in the air, brought it down on the back of a hand, and sighed.

"Okay, you're in charge, capt'n. Funny, that nickel never failed me before. Well, fair enough. I won when we flew the Fireball."

They walked the sprawling circumference of the plane, hoisted themselves in, pulled the heavy door shut with an echoing thud, and made their way down the unlit interior. Jack entered the cockpit door and climbed right, Ned followed, they did the checklist, and started the engines.

"Our tower boys must be halfway stateside," Jack mused. "Chinese replacements are coming in, and we're on our own to get outta here, that's for sure."

Ned soon had them bumping down the rutted runway, steering into the wind . . . always into the wind, his mind rolled back reciting stuff he taught others.

They were gaining the fence, powering into liftoff, when suddenly Jack shouted.

"Son of a gun! You see that?"

Ned saw. A single-engine P-40 coming up their tail like a firecracker, with a Chinese pilot at the controls. Guy must've hopped in and revved up while they were doing the preflight check-over. If they lift ol' Dumbo too fast now, they're going to collide midair.

Ned nudged the nose and banked wide of the cliffs, so low he glimpsed a terrified farmer making a dive for his life headfirst from a dike into cold water. They strafed by just missing him. The pilot of the smaller plane gave it the throttle and overpassed in a spew of smoke.

"*SON OF A . . . !*" Jack shook his fist as the stuntman pulled away.

Ned leaned back on the yoke, hit the throttle, got the gears up, and moved the plane fast into the climb. Gotta hand it to the Commando today, he's showing his pep.

"What do you say, Ned? Let's teach that little squirt a lesson."

"What've you got in mind?"

"A '40 can dive, but I bet he can't outrun us if we take him by surprise. Let's open this baby up all the way."

"He's just showing off," Ned shrugged.

"Well then, show him what we're made of. You with me?"

Ned pushed the throttle again. Tweaked the flaps and rudders, threw a glance at his RPMs, and peeled out. Giving the smaller plane a certain berth, still he cut it close enough to let him know who's boss, then left him behind in the plasm of their trail.

Jack cut loose with a long, obscenely guttural whistle.

"You made one scalded dog outta that pup. Rubbed his nose in it."

Ned's mouth quirked to the side for a hair of a second.

"Couldn't have done it much better myself," Jack allowed generously. "Let him be the last plane out of here. Doesn't bother me, flying the last of Uncle Chump's '46s to Shanghai. Which I am fully convinced we're doing. Just don't want to be the last plane out, period. From anyplace. You get my drift?"

"I feel kinda bad for that farmer," Ned said.

"That comes from you havin' a soft Carolina heart, same as mine, so don't cry about it. That hot-rodder could've got us all killed. They'd be months pecking our body parts out of the cliff face with a toothpick."

Lifting into the low-lying clouds, they had the plane trimmed up and climbing, heading east towards the sun. Assuming the existence of a sun. Some things you take on faith a lot of the time. Twenty minutes out, at 9000ft, they switched to autopilot. Jack leaned back, pulled his nickel from his jacket and lay it buffalo side up on the pedestal.

"May Lady Luck lead the way," he intoned solemnly.

"Put your luck away, Mayer, we'll get home. Just have to backtrack across China first. Get this plane to some hotshot over in Shanghai. Who maybe forgot his fountain pen last time he was on board. Little round-trip jaunt the width of Asia, and he'll set up his fine new desk."

"That's telling it, Thomas. Hang around me, soon you'll be as cynical as I am. Tell you the truth, I expected a load of Chiang's boys today. With our luck, they'd all have fountain pens. Can you see them jabbin' holes around the inside of the plane? Bam, bam! You and me, we'd be around to the knell of doom, and not done moving those guys or plugging up pen holes. This last plane could've been our unluckiest yet."

"Well, I'll trust myself more than Lady Luck any day," Ned observed.

"You say? Well, you were always the one with the confidence. What I hear, no one ever saw you lose your nerve. Not once."

Ned gathered in the remark. Couldn't say he was a man with nerve. Just doing a job.

Jack, who never stayed quiet for long, pressed on.

"Not even that night we flew into Shanghai, and lost our landing lights. Did we have some kinda time setting that baby down. Could'a ditched in the harbor. I was sweatin' blood."

"No big deal." Ned replied even-humored. "Another one to write up."

"Well, let me tell you, as soon as I get myself reinstalled in some honest line of work back home, you better believe I'm not going to miss these Chinese runways, lights or no lights. Wet sand gummin' up the works till you think you're jockeying around the high sky in some whirly-bird cement mixer."

"I won't miss the sand," Ned agreed.

"You ever sweat a flight? I mean serious perspiration. Even once?"

"Sure," Ned answered. "Plenty of times."

"You pullin' my leg?"

"Couple weeks back I was looking for a place to set my plane down, right out here in the middle of China. Radio came on and scrubbed my mission. I was past the no-return, then my main and auxiliary tanks red- lined. Headed to Kunming, come to think."

"Kunming? Stop! Don't say any more. It's always bad luck to dwell on trouble nearby."

"I'm not dwelling. I was glad to see Hankow again. Even when I thought Langford was going to tear the bars off my shoulder. Then he keeps shouting about staying in China and doing good. Some newfangle Orient idea. I should've kept going and got refueled. Orders or not."

"Well, going down in a paddy is one brutal event." Jack shut both eyes, the better to visualize. "Rips your wings off. Splits your fuselage, cracks it open like an egg. Cockpit glass shatters around you like knives, and there goes the last of your human form . . . .What am I talking about? We've all seen this stuff. I'm as good as begging disaster to strike. Let's stay positive, okay? What d'ya say?"

"Positive it is," Ned concurred.

"They better have fueled this ship up good at Kunming," Jack chomped. "Anybody shorted us, I am going to make it my personal business and mission in life to track him down like a hound, and punch out his lights with my bare hands. Near to three thousand miles we gotta fly in twenty-four hours. Do those vampires care? Not on your life. Coming at you with their fangs bared, suckin' the juice outta your lines, smacking and licking their lips. Smiling like ghouls. You get a wire from me at Leavenworth, you know what I'm in the brig for. Send a hacksaw."

"I'll mail it in a turkey," Ned said. "For the holidays."

Jack peered out the wide solarium of the front, then down the side at shoe level.

"Where'd you get in trouble? Relative to where we are about now?"

"When?"

"The day you redlined the tanks, for crying out loud."

"Little farther than here," Ned answered reassuringly.

"Just how far were you before you crossed the Rubicon, so to speak?"

"Maybe two-thirds of the way, I'm not sure."

They settled in. The flight was going to be long. China could've been banked in from Ghenghis Khan's Wall to the South China Sea. In half an hour, Ned figured the tedium had put his hyper-alert copilot to sleep. Then abruptly Jack tacked into a new line of discourse.

"Have you thought about how long we gotta hang around Shanghai, waiting for berths on a ship?"

"Hadn't thought about it unduly," Ned allowed.

"Well, I have. They say it could be weeks. And I'm planning to put that time to profit." He sat straighter, reached into a jacket pocket, pulled up a narrow money pouch, and unzipped it. Ned's eye took in the small palm-size brown bottle he extracted.

"My secret invention. Which I figure to perfect right here in good old China, just in time for us to ship out. Say you're interested, Ned, I'll bring you into the action on the ground floor. Believe me, I'm not making this offer to just anyone.

"Now this little vial may not look like much." He held the bottle to the light between right thumb and fingers. "But I am here to tell you, it will do the job every time, and you can go to the bank on it. Gives a man confidence and hope when he needs it most. My own unique creation based on ancient and revered Chinese recipes and secret ingredients. Let the word get out, old man, you better stand back for the stampede."

"What've you got in there, Mayer? Bat droppings?'

"Ye of little faith! Ye who shall regard the contents herein, in days to come, as a miracle of formulaic genius. Molten and piercing as dagger blades in a flaming forge. Stands up under any conditions of nature or adversity. Like July fireworks on a red-hot grill out in the equatorial sun. An unprecedented concept of pungency, potency—"

"Come on, Mayer, Asia's full of this junk. You see guys selling it on every street corner."

"Not like what I put in this bottle."

"Well, good luck to you, my friend. You're going to need that nickel after all."

"You're not catching on, Thomas. This invention isn't for the Chinese marketplace. It's for Americans. On board troop carrier ships like we're sailing home on."

"What use would that trash be to men on a ship?"

"Could save their lives."

"Right-o, Mayer. In the event we run across a school of mermaids?"

"My friend, you wound me. It's for when we meet up with a school of sharks. Listen here, Ned. No, just smell this stuff."

"No thanks."

"Smelling is believing. Now everyone knows sharks have keen noses. Or whatever it is they use to smell with under those beady little nostrils of theirs. And what makes a shark go starkraving crazy? The smell of human flesh. Say they take a nibble on some guy's ear, and there's a drop of blood. That blood is gonna set off a feeding frenzy, till first thing you know, they're chompin' everybody's leg off, then their arms, then our heads, and the rest of us right and left, till we're just a mishmash of hamburger gommed up, out there in the water. They're cruisin' around us with their fins and fangs, laughing their evil laugh and spreading terror through the ranks. Justifiably. I'd be scared if a herd of sharks came galloping around me."

"I would, too, but what's that got to do with your bottle?"

"That's what I'm trying to tell you. I invented a shark repellant. What were you thinking it was? You see here my own secret blend of the hottest Chinese peppers distilled in a piquant base of cabbage and certain oily fuel components that make the stuff stick to trouser legs like tar, and penetrate fabrics—wait, don't tell that part. I don't want anyone stealing my idea."

"Shark repellant." Ned rolled his eyes.

"I hear Uncle Sam's got his egghead boys working the idea back in the labs. But I guarantee you, no formula can hold a candle to what you see right here before your eyes. I just need time to mix and bottle enough to sell when we get on board. There could be a run on this stuff. I want to have plenty, so when the ship goes down—"

"Mayer, our ship isn't going down."

"Sure of that, are you? Remember the Titanic. Wasn't even a war on. How many of those sad souls do you suppose got eaten up by packs of wild roving sharks, cruising around looking for trouble? Moving in like wolves. Now this stuff is so easy to use a child can do it. Fits in the pocket of any life jacket. You just apply some to the leg area as you leap off the deck, maybe the cuffs of your shirt for extra insurance, then swim like mad, far enough so the ship doesn't suck you under. You know how easy it is to get pulled down the vortex of a doomed ship."

"Shark repellant," Ned repeats.

"As you live and breathe." The inventor falls silent, leans back, plants a boot up on the dash, and folds his arms. Ned flings a glance through the window at the props and runs his eyes over the dials. Through occasional breaks in the clouds, he could make out the Yangtze River below, spinning a thread-thin tangle across the endless "good earth" of China. By and by, Jack's voice again broke into the steady ear-pounding roar of the engines.

"What're you going to remember about this place, Ned? I mean about the Hump. I mean later, after you get home."

Ned shrugged. "Same thing everyone remembers, I reckon."

"What I'm saying, is what stands out in your mind?"

"Well, they're flying General Marshall over in a couple weeks.

They say no more dollars or equipment to keep U.S. troops in a foreign environment, ever again. Uncle Sam's done fighting wars for good, and they're pulling our main—"

"No, Ned, no. Nooooo. I'm asking you. There'll always be another big muckety-muck getting sent in and out of somewhere, doesn't matter if you and me die young, or die old."

Jack brought down his boot and focused a look to the pilot side. "Supposing a long time from now, maybe you're old as Methuselah. Hair's turned white, you got liver spots on your hands. Knees gone soft. Ol' rockin' chair got you. Then one day out of the blue, somebody puts the question. '*Sir, is it true? That you flew those heavy old planes way up there over the Top of the World? Why, that must've been a sight, Sir. What did you see up in that big sky?*'"

"Backseat of my copilot's pants when we got flung around," Ned quipped. "Icebergs crashing against the windows between lightning bolts."

"That all you can say?"

"Well, my nails didn't always look real healthy."

"Who cares about our blue hangnails and all the craziness? We could have been flying over Mars! No one else ever did this before." Jack's voice mounted above the roar of the engines. "They wanna know what *you* saw. If they want statistics, they'll read a history book."

"Ha! You can forget that one, pal. The Hump won't make the history books."

"Why not?"

Ned cogitated briefly, then shrugged. "No one would want to hear. Probably wouldn't believe it anyway."

"Well, supposing they do want to hear, Ned Thomas? So don't be a schmuck. You can bet your life somebody's going to want to know someday. And they are going to believe it."

* * *

Major Langford's banquet gets cancelled in the departure rush of Shanghai. The last day of November, they're checking the men through to board. Pilots, mechanics, navigators, engineers, GI's, all swarm down the seedy city docks with their bags and gear, to find their place in the lines. *Hurry up and wait.* The thudding mantra of military life chomps down on the Tenth. It takes a few days to get every man assigned to a berth, settled, and accounted for. Till at last they're pulling from the harbor, looking ahead to the three-week crossing to Seattle.

From the other side of the planet, far away in Washington, President Truman sends his reassurances to a proud Generalissimo. America recognizes the legitimacy of Nationalist governance and hopes for unity in his strife-torn nation. Free people everywhere look to the advent of democracy in China, as opposing factions come into dialog and clearer thinking. Gazing out the window of the Oval Office, does he breathe a prayer of thanks for the tired shoulders of General Marshall to help carry this burden of diplomacy? Who'd blame him?

Deep within China, Chiang Kai-shek experiences a rare flicker of hope and expectancy. Surveying his losses and gains, he is not disconsolate. The conflict with his tricky former comrade Mao is no bed of roses. Still, he retains the blessing and good will of his dear American friends. Heart-weary of war, a new dawn seems to break on the day of December 16, when he and his cavalcade spill into Peking for the first time since Japan invaded his homeland. Entering the old Forbidden City, which an American pilot recently discovered silent and nearly deserted, swamped by adoring Chinese mobs who surge around him, he sheds tears as he exhorts them to join together to rebuild the nation. 55

Out on the wintery gray Pacific of mid-December, the Victory Ships plow their rough and foaming course. The hours tick by, the days, one dragging on another, in the slow pace of shipboard life. In quiet wonder, past words, the men look east toward home.

No vessels get lost. Nothing untoward befalls to worry a man. You almost imagine General Marshall's plane passing somewhere over your head in the opposite direction, as he rushes westward to China, where he is, for sure, going to set things right at last, with Chiang.

Ned finds he does not take to sailing. It's the motion. His top bunk in a 4-tier rack, green eggs on the menu, the vertigo-inducing roll of the deck, do not foster an appetite. A camel ride would beat this. Given the choice, he'd have flown, but he bucks up and thinks about home. It could've been another lifetime when he and Neva Rae said goodbye.

Now he'll be starting his life over. A complex proposition. Keeps his mind busy.

The way he sees it, he'll quit flying and stride out to make his mark. Drop the silver wings in the back of a drawer and forget them. Why does that part sound a little sad?

Standing out on the wave-swept deck of the Victory Ship, from a pocket he pulls a weighty, solid round medal of dark brass with Chinese lettering. Generalissimo Chiang sent it to men like himself who'd stayed after the war and flown his soldiers, many of them the same pilots who'd risked their lives over the Hump.

Sir, is it true what they say? What did you see up in that big sky?

He drops Chiang's medal back into his pocket, leans against the railing, and looks out on the spray of a roiling slate-gray ocean. Hears the deep-voiced, groaning labor of the ship as it sleds the waves. Let someone just go out there and split the whole sea wide apart like old Moses at the waters—well, what would a human man behold? Mountains that march along somewhere in the deep? Maybe as far below him today as he'd flown above, in that strange land where the tallest peaks in the universe pitch one into another, to form the Roof of the World.

Truth to tell, he'd tried not to think about it. Focused on the mission, on the flying.

Yet sooner or later, above the vast Himalayas comes a break like a parting of the seas. Masses of air and wind, water and weather, suddenly roll aside to open a view of a blue-lapis heaven. Underneath emerge banks of cloud and snow-shrouded peaks. A pilot could think he was sailing a vast high seaway rocked by streams and currents, tossed by virulent undertows, where mountains ripple into infinity, in contours of irregular undulating waves, their ragged white points breaking through the mists in a colossus of whitecaps. Now look again, this time out to the rim of the world—for there trek the super-giants no fool would try to fly over.

The Hump. Silver peaks against a turquoise sky. Black rock face and swirling snow. The harsh walls of the treacherous unbreached heights. Folded in among them, the long roaring rivers that first pierced earth in distant eons before the footprint of man. You fly on till an old monastery frames itself like a painting in your field of vision. Then for a spit of time, you shift a bit higher in your perch and exhale. Because you've come through another crossing alive. Exhilarating? Who can answer that one? You're not signing up to do it again tomorrow.

Yet in *retrospect*—let's get a lot of retrospect here—you did a job right, for all the right reasons. Part of a great team. Men at war. You saw it through.

PEEL BACK THE SKY

IN TRIBUTE TO THE MONOLITHIC AIRLIFT THAT CROSSED THE Top of the World, Theodore White and Annalee Jacoby collaborated to write *Thunder Out of China*, in 1946. They certainly don't skimp on the drama in their portrait of the pilots who flew the Hump:

"The Hump drove men mad, killed them, sent them back to America wasted with tropical fevers and broken for the rest of their lives. Some of the boys called it The Skyway to Hell; it was certainly the most dangerous, terrifying, barbarous aerial transport run in the world."[56]

Broken? Terrified? Mad? Ned laughs. "I personally don't know anyone who went mad. Last time I checked, I seemed to be doing all right myself. Allowing that sort of thing may happen sometimes, still the men I met up with mostly looked okay to me. That's not to say the Hump was a pleasure trip, but this suff added on . . . understand this now . . . we just went out, flew the airplane, came back, flew again, and that's your job. I don't know where authors come up with these assessments, but in my estimation they're overblown. Sure, there were men who refused to fly. As in all wars, some return incapacitated mentally or physically. I can't particularly name any, but there are bound to be a few.

"Let me tell you something. I can recall sitting in that cockpit, I'm a young fellow, and I know I've got a lot of flying ahead. I've looked at my watch and decide I'm not going to look at it again. So I scan my panel. I look out the windows. Check my dials. Then my engines and fuel. Then the dials again. I inspect my fingernails. Till I figure finally some time's gone by, so I guess I'll look at that watch. And two minutes have passed."

Slugging through the nerve-building tedium, this, too, formed part of the dark, dragging hours of instrument flying between assorted hair-raising emergencies, that might be dire but not inevitably fatal. Could the authors only have glimpsed something vastly more positive among the men who emerged from the Hump experience—their enhanced life skills, kindness, courage, and spirit of camaraderie that came from teaching each other the ropes flying the Rockpile. To keep the memories alive, they formed the Hump Pilots' Association, which had them in touch through 2005, when the group disbanded. By then, so many had flown out to a Higher Sky.

* * *

From time to time, a downed Hump crew would be rescued from the Burmese jungle. One recovery story involves the notable American news correspondent, Eric Sevareid, who as a passenger in a C-46 that lost an engine, parachuted down with some twenty other passengers. They were located by a daredevil pilot named Blackie Porter, who swooped over the trees to drop in supplies till they could be evacuated. All survived, save Porter alone, for it proved one of his last missions when his own plane went down in the Hump shortly afterward.

For men who crashed into the pitiless peaks of the Himalayan spine, fatalities ran one hundred percent. They said if you could conquer the Hump, you could conquer anything. But one lapse, a split-second misjudgment, could ditch you. Pilots crossed on courage and superb flying, their aim not heroics, but to get down the slope, bring in their overworked planes, and stay alive to fly again. Seldom were downed aircraft even sighted in the depths of the crevasse. The lost crews of the Aluminum Trail could have been sucked under a polar cap.

Fast forward to the year 2000, and an Associated Press release: China's Foreign Ministry has been in touch with the Pentagon. Two crash sites have been discovered, one in Tibet, involving a Hump C-46 carrying four crew members. The Chinese promise to assist the U.S. in search efforts. The following year, 2001, a mountain climber from Phoenix, Arizona, Clayton Kuhles, embarks on a climbing expedition to Nepal, and on a hunch, diverts into Burma, does some searching, and amazingly, finds a C-47 on a mountainside near the India border.

Following his discovery, Kuhles will return each year at his own expense to search for downed Hump aircraft. His dedication leads to the identification of several planes, and brings word to families who'd lost hope of hearing the fates of their loved ones. Using engine fragments, anything with a serial number, respecting bodily remains and tracking data through military records, Kuhles painstakingly pieces together the final stories of several crews. His work gains new attention for the Hump War and its lost. Then in June, 2012, India agrees to allow American military search teams to enter the range and continue searching for remains.

On March 30, 2013, the flag is raised over the Flying Tigers Heritage Park and Museum in Guilin, China, to commemorate every American pilot who flew in the theater in World War II.

"The Chinese have never forgotten what these brave men and women did for them during the dark days of World War Two. The Chinese do not distinguish between flying forces but rather consider all airmen who served in China to be Flying Tigers— Fighter Pilots, Bomber Pilots, Transport Pilots, and ground crews. . . . With Courage, Honor, and Valor, impossible odds were overcome." In tribute to their heroism, the address recognizes all who flew in Asia as "Flying Tigers" but each force is praised for its unique mission and dangers distinctive to their commands. At last, the men of the Hump—what Clayton Kuhles calls their "suicide mission"—are receiving the honor and recognition they earned so dear.

* * *

Early Summer, 1972

What brings Colonel Ned Thomas, this bright clear day, to Soesterberg Air Base, Holland, home of the U.S. 32nd Tactical Fighter Squadron for NATO? More than a few pairs of eyes follow his chauffeured car as it pulls through the gate. The salutes carry good will. There's been talk about this American Air Attache who'd recently been knighted by the Dutch.

Ned alights at Administration, where he's greeted by base commander, Colonel "Hoot" Gibson, the one man hereabouts who may have the story behind the talk of this mature Command Pilot, who'd rejoined the armed services with the birth of the U.S. Air Force in 1947, whose career has taken him to Japan, then to war in Korea, on to the Philippines, and later into diplomatic work, first to the American Embassy in Paris, now the Embassy at The Hague.

Gibson's handshake is warm. This diplomat and fellow officer has done a lot to get some sleek new fighter jets headed to his base, here in Holland. He knows about the trust Ned has built with officials at The Hague over the past three years, the good spirit he generates, the high marks he's earned for his skill in working with people. There's been talk of his advocacy for an equal opportunity initiative, and of his gaining local support for the U.S. Prisoner of War program.

But how about those new jets?

The Netherlands, a NATO country with its own treaty and defense commitments, wishes to modernize and replace her fleet of aging fighter-bombers by bringing a capable, small fighter into their military. And as it happens, U.S. General Dynamics stands ready to roll a beautiful aircraft off the line, the single-engine F-16 Fighting Falcon, if it can be sold.

Competition arose, naturally. England hopes to sell a twin-engine of their design. France and other nations have productions. With so much at stake, it hardly could've been worse when the U.S. nearly fumbled the ball, but the head at MAAG—U.S. Military Assistance Advisory Group—gets suddenly recalled to Washington, leaving a potentially lethal negotiating gap.

Ned's grasp of his own country's needs, his knowledge of aviation, combined with his familiarity with the Netherlands Air Force and senior Dutch officials, enables him to step into the breach, cover vital details in the talks, and provide insight for his ambassador and industry reps. He proves instrumental in helping win a sale—for in fact, the business snowballs, and other NATO countries line up to purchase the F-16. You never know where the past will lead, in this case, a pilot's care to learn the complex workings of many aircraft, and his dedication to building this knowledge into a lifelong storehouse. Naval Attache Captain Joseph A. Grace files an official report regarding Ned's "guidance beyond that which would be expected of the normal attache. (In) the absence of the Head AF MAAG, Colonel Thomas alone carried the military support to the Ambassador's tremendous efforts to promote the sale of a U.S. fighter to the Dutch—and thus to Europe." [57]

The American Ambassador, J. William Middendorf confirms the reliance on Ned: "his professional knowledge of the aircraft which we are frequently called upon to discuss with Netherlands officials, particularly to promote fighter aircraft and equipment of American origin for use by the Netherlands." [58] In a letter to the Secretary of the Air Force, Robert C. Seamens (July 25, 1972), the Ambassador names Colonel Ned Thomas as one of the finest Air Force officers he has ever known. "His ability, combined with his pleasant and affable demeanor, ready wit, courtesy, and obvious sincerity, brought him respect and confidence at all levels of command, as well as in the business and professional communities of this country. I have frequently been advised by leading representatives of the high esteem in which he was held. I have relied heavily on his advise and judgment. His thorough knowledge has provided the basis for decisions regarding Air Force matters. His competent efforts have contributed significantly toward the sale of U.S. aircraft and aviation equipment. The job Colonel Thomas did here was so superior that I thought it important for me to personally advise you of that fact." [59]

Ned and his wife Neva Rae, notes Captain Grace "make a top notch Air Force and Attache team." For Neva Rae is there, a gracious helpmeet and hostess, in every way part of the "team" to build friendship and good will. As their tour draws to an end, the Dutch give special celebrations honoring them both. "Why are you leaving when we have come to love you?" they ask. The country inducts Ned into the royal order of chivalry bearing the ruling family's name, the Ordre van Oranje-Nassau, with the grade of Commander, its wide blue and orange chest ribbon, medallion with crown, and shield with crossed swords.

This is an old monarchy. Endearingly Neva Rae remembers their petite Queen Juliana. "She seemed to be a very quiet person. I got the impression she was rather shy. I'd see her in church, sitting up front all alone in her pew. She was tiny, and her feet didn't reach the floor. I can just see her there on that bench, swinging her little feet back and forth." Prince Bernhard of the German House of Hanover, often pulls Ned aside for a friendly word, and the royal family receive the couple at the palace in Amsterdam, happy moments of a warmhearted nation, where even Princess Beatrix, heir to the Dutch throne, wheels about Amsterdam on her bicycle.

At Soesterberg, Ned and Hoot emerge, turn purposeful steps toward the flight line, and come to a stop before a familiar scene. The present fleet of U.S. F-4 Phantom jets.

As a diplomat of forty-eight, Ned has left the cockpit to others for awhile. Not that older pilots forget, and odd times, there comes a tug. He misses having his hands at the controls.

His eyes take in a handsome sight. Strung along the tarmac, the big Phantom fighters tuck dual jet engines under their wings, each packing a volcano of power, their curved fuselages furling to 62 ft plus, with a refueling probe to extend the nose like a rapier.

Following a familiar pattern, Ned circles the plane's perimeter. His eyes scan the double-cockpit canopy, move up the wings, down the streamlined silhouette of the tail. An aircraft built to move, if ever one was.

Hoot's voice breaks in. "Ready to take her up, Colonel?"

Ned nods. To experience the Phantom. He's ready.

"We pulled you a six-footer suit and headgear over at Operations. Only thing, let's make sure you know the ejection system. Our Martin-Baker mockup will have you a pro in no time." He lands a hearty slap on his companion's back. "Ned, fly the F-4, and you'll wonder why you bothered with those dinosaurs in the old days. I guarantee you won't regret it."

"Hoot, you know I don't waste time regretting."

"Laughing are you? Just don't tell me I oversold this one. Ever have problems with motion? What am I saying, an old pro like you? I sent around for Captain Baynor, he'll be your copilot. Now for that ejection seat—"

In ten minutes, he's suited helmet to boots. An enthusiastic cluster of pilots and mechanics gathers near to see the Colonel learn the ropes.

Things sure change. To think pilots once flew the world in threadbare khakis and bomber jackets. Learned the kid-stuff in prehistoric Link Trainers. Went down kinda hard if their medieval parachute didn't open right, maybe with a knife clutched Tarzan-style between their human teeth. Stone-Age all-the-way.

He settles into the futuristic, high-back Martin-Baker, and has it down to an A-plus. The seat isn't live, but in the plane it is, in the event they encounter some worst-case deal and have to eject out there over the North Sea. You don't look for trouble, just keep prepared for what can come along, same as you've always done. Flying hasn't changed in some respects, come to think.

The men watch him cross the ramp. Glimpse as he steps over the wing.

Old Commando, modern Phantom, the cockpit is familiar as home. Ned's got his right hand on the stick, left on the throttle, feet on the rudders. A minute later, they're bumping down the runway. There comes the sudden pressure, acceleration, the nose lifts, gears retract, they move out in the streak of a rocket.

You work a lifetime for this. To feel the power and performance of flight, and it all comes together like a dream.

High above the flat, enduring polders of this land reclaimed from water, flashing away toward the cold North Sea, the Phantom peels back the sky and outruns its roar in long spumes of vapor. Until all anyone sees is a glint of silver in the sun, as the plane punches into the vault of blue, and leaves silence.

NOTES

CHAPTER 1 SOME HIGH HARD ROCKS

1 *Life Magazine*. July 20, 1942.
2 "The Flying Humpty Dumpties," *The Reader's Digest Magazine*, May 1944. p. 42
3 Lin Yutang, *My Country and My People*, pp. 368 -399.
4 Lin, p. 428.
5 Further reading on Stilwell's behavior toward Chiang: Jay Taylor, *The Generalissimo, Chiang Kai-shek and the Struggle for Modern China*, pp. 235-6; & 257-259.
6 Lou Thole, *Cotton, Tobacco, Peanuts and Pilots*, p. 13
7 Thole, p. 16
8 William H. Tunner, *Over the Hump*, p. 6
9 Pamphlet *"The First Twenty Questions,"* referenced by Thole, p. 12.

CHAPTER 2 "NOOKS"

10 Theodore White, Annalee Jacoby, *Thunder Out of China*, p. 154.
11 William Koenig, *Over the Hump, airlift to China*, p. 117.
12 Koenig, p. 116.

CHAPTER 3 "PRIVATE PROPERTY: NED THOMAS"

13 William H. Tunner, p. 63.
14 Tunner, p. 135.
15 Tunner, pp. 28-30.
16 Tunner, p. 15 & p. 41.

CHAPTER 4 FLYING THE FIREBALL

17 Jeff Ethell and Don Downie, *Flying the Hump*, p. 27.
18 Otha Spencer, *Flying the Hump, Memories of an Air War*, p. 41-42.
19 Spencer, p. 42.
20 Tunner, p.43 & p. 63.
21 Tunner, p. 88.
22 Tunner, p. 114.
23 Tunner, p. 86.
24 Tunner, p. 87.
25 Tunner, p. 58.
26 Tunner, p. 113.

27 Tunner, p. 91.
28 Spencer, p. 25.
29 Tunner, p. 74.
30 Ethell and Downie, p. 121.
31 Chick Marrs Quinn, "Hump," *Operations of the Air Transport Command, December 1942-August 1945*. Another summary of Quinn's statistics see Spencer, Notes, p 183, note 1.

CHAPTER 5 HOUSE OF SNOW

32 Tunner, p. 101 - 104
33 Tunner, p. 103.
34 Tunner, p. 110.
35 Tunner, p. 105.
36 Tunner, p. 106.
37 Tunner, p. 146.
38 Tunner, p. 73.
39 Ethell and Downie, p. 11.
40 Tunner, p. 74.
41 Joseph Stilwell, "History of the India-China Division, Air Transport Command, December 1942 to October, 1945," *China-Burma-India, Theater History, 1941-45*, pp. 15-16. Quoted in Otha Spencer, pp. 50 - 51.
42 Tunner, p. 119.
43 Tunner, p. 45 & 62.

CHAPTER 6 AN EPIC OF THE WAR

44 Spencer, p. 164, quoting from *The Pacific: Matterhorn to Nagasaki*, by Craven and Cate, pp. 170-171.
45 Tunner, p. 115.
46 Tunner, p. 131.
47 White and Jacoby, p. 155.
48 Tunner, p. 133.
49 Tunner, p. 127.

CHAPTER 7 "IF THE PLANES WERE READY, WE FLEW"

50 Some of Chiang's hoped reforms are discussed in Jay Taylor, p. 305.
51 White and Jacoby, p. 164.
52 Tunner, p. 10.
53 Tunner, p. 130.

54 Tunner, p. 8 - 9; 59
55 Taylor, p. 329; I thank the author for confirmation of this date.

Epilogue: Peel Back the Sky

56 White and Jacoby, p. 154.
57 U.S. Military Document, April, 1972
58 Hon. J. William Middendorf: letter to Major General Rockly Triantafellu at ACS Intelligence HQ, Washington, 2/10/71. Document in author's possession.
59 Letter to AF Secretary Seamens, and others here, reside in Col. Thomas' military records

Bibliography

Ethell, Jeff and Downie, Don. *Flying the Hump.* St. Paul, Minnesota: Motorbooks, 2002.

Koenig, William J. *Over the Hump: Airlift to China.* New York. Ballantine Books Inc., 1972.

Hammond's Historical Atlas. Maplewood, N.J. C.S. Hammond & Co, 1963.

Lin, Yutang. *My Country and My People.* Bombay, Madras. Jaico Publishing House, 1935-39.

Quinn, Chick Marrs. *The Aluminum Trail. How and Where They Died.* Florida. Copyright 1989.

Sevareid, Eric. "The Flying Humpty Dumpties." *Reader's Digest*, May 1944, pp. 41-43.

Spencer, Otha C. *Flying the Hump. Memories of an Air War.* College Station, Texas. Texas A & M University Press, 1992.

Taylor, Jay. *The Generalissimo, Chiang Kai-shek and the Struggle for Modern China.* Cambridge, MA. Belknap, 2009.

Thole, Lou. *Cotton, Tobacco, Peanuts and Pilots: The story of the 63rd Flying Training Detachment.* 2010

Tunner, William H. *Over the Hump.* (Reprint of 1964 book by Gen. Tunner and Booton Herndon) Scott AFB, Illinois. Military Airlift Command, 1991.

Vaughn, Don. "Lost in the Himalayas." *Military Officer*, September 2009, 93-99+.

Watry, Charles A. *Washout! The Aviation Cadet Story.* Carlsbad, California, California Aero Press, 1983.

White, Theodore H., and Jacoby, Annalee. *Thunder out of China.* New York: William Sloane Associates, Inc., 1946.

INDEX

Air Transport Command (ATC), 4, 95

Aluminum Trail, The (C. Mars Quinn), 95

American Volunteer Group (Flying Tigers), 10

anoxemia (anoxia) 102

Arnold, General Henry "Hap," 5, 14, 32-33, 117, 140, 143

Artificial Horizon 1920s (Gyro horizon) 122, 126

Asia Armistice, xvi

Assam Valley (India), 3, 91, 105

AT-10 Wichita (by Beechcraft), 36, 37, 46, 47

ATC (Air Transport Command), 6, 95

Australia, 3

B-17 Flying Fortress, 11, 53, 104

B-24 Liberator, 11, 104

B-25 Mitchell, 10, 36, 53

B-29 Superfortress, 31, 78-80

Baer Field (Fort Wayne, Indiana), 69, 70

Bataan, Philippines, 3

Battle of Britain, 10

Battle of the Bulge, 56

Battle of the Coral Sea, 3

Battle of the Philippine Sea, 56

Bergstrom Army Airfield (Austin, TX), 61, 63

Berlin Airlift, 159

Bishop, General Cleo, 21

Bonner, Major Thomas W., 22

Brahmaputra River, 91, 105, 118

Brereton, General Louis H., 14

British Malaya, 3

BT-13, Vultee, 29

Burma, 3, 19, 34, 84, 107, 117

Burma Road, 3,19, 118

C-46 Curtiss Commando, 63-67, 80-82, 119, 120

C-47 (DC-3 Dakota), 14, 53-56, 65, 120

C-54, 53, 54

C-109 & C-87, 127

Casablanca Conference (Jan. 14-24, 1943), 14

CATF (China Air Task Force), 95, 115-117

CBI (China-Burma-India) Theater, 58, 73, 89

Chabua, India, 4

Chamberlain, Neville, 44

Charlie Route, 104

Chennault, General Claire Lee, 3, 10, 14, 18-20, 27, 31-33, 57, 84, 114-118, 140-141

Chiang Kai-shek, xvi, 2, 3, 10, 13-18, 20, 31, 32, 35, 57-58, 79, 146, 148, 157, 160, 174

Chiang, Madame May-long Soong, 15, 18-19, 82, 114

Chihkiang (Zhijiang) 84, 149

Chungking, 3, 31

Churchill, Winston, 10, 14, 17, 31, 44, 56, 142, 160

CIS (Central Instructors School) Randolph Field, 38, 46-47

CNAC /Chinese National Aviation Corporation, 14, 34, 95, 114

189

Cochran Field (Macon, GA), 28
College Training Detachment (CTD), 12
Comfort women, 151
Coral Sea, Battle of, 3

D-Day, Normandy, 31, 56
Dinjan, India, 4, 91-92, 98
Doolittle, James, 10 - 11, 28
Douglas Field (Georgia), 20-27
Downie, Don, F.O., 74, 94
"Dragon Deaths," 150
Duquesne University, 12

E6B "whiz wheel, "124-126
E-Route /Easy/ Lower /Southern Hump), 103-104, 106
Eisenhower, General Dwight David, 31
Enola Gay (B-29), 78, 145

F-4 Phantom, 183-184
F-16 Fighting Falcon (General Dynamics), 181-182
Fireball Route, 73-77, 80-83,
Flying Tigers Heritage Park, Guilin, 179
Fourteenth Air Force, 10, 117
Fourteenth Volunteer Group/ Fourteenth Bombardment Squadron (Flying Tigers), 115

George, General Harold, 73, 85, 141
Gibson, "Hoot," Col., 180, 183
Grace, Captain Joseph A., 181, 182
Guadalcanal, 3
Guam, 3, 56

Hankow, 2, 43, 152, 153
Hardin, General Thomas O., 59
Himalaya Mountains, 2 - 5
Hiroshima, 145
Hitler, Adolf, 43, 44, 78
Hoag, Gen. Earl, Commander India-China Wing, 32-33
Hornet, USS, 10
Hurley, General Patrick, 163

Ichigo (Japanese offensive), 31, 32, 34
Indonesia, 3
Iwo Jima, 56-7

Jacoby, Annalee, 31, 142, 159, 177
"Jing Bao," 108
Juliana, Queen & Prince Bernhard, Netherlands, 182

Kuhles, Clayton, 179, 180
Kunming, China, 91, 84, 103, 136-137, 164-165
Kuomintang, 15

Langford, Major John V., xxv, 92, 149, 152, 154, 174
Ledo Road, 19, 91, 105, 118
Lend Lease, 5, 17
Light Line, 9
Lin Yutang, 15-16
Lindbergh, Charles, 9
Link Trainer, 28
LORAN (Long Range Navigation), 99
Los Alamos, NM, 141
Luce, Clare Booth, 3

MacArthur, General Douglas, 3,

56, 146

Malden Army Airfield (Missouri), 48-50

Manchuria, 2

Mao Tse-Tung, 15, 94, 146, 162

Marshall, General George Catlett, future Secretary of State, 34, 58, 117, 163, 174

MATS (Military Air Transport), 159

Maxwell Field, Montgomery, Ala., 13

Merrill's Marauders, 19

Middendorf, Ambassador J. William, 181-182

Midway, Battle of, 11

Moody Army Airfield (Valdosta, GA), 31, 35

Mountbatten, British Vice Admiral Lord Louis, 17, 119

Murrow, Edward R., 43

Mussolini, Benito, 78

Myitkyina, Burma, 4, 19, 35, 104, 118

Nagasaki, 145

Naiden, General Earl, 14

Nanking, 2, 43

Nationalist China, 2 - 5, 13, 31, 32, 148

New Deal, 43

Nimitz, Admiral Chester, 56

Okinawa, 57, 78

Operation Grubworm, 58

Operation Matterhorn, 78, 80

Orient Project, xxv, 146-147

Pearl Harbor, 2, 21, 45, 115

Peking, 2, 154-156

Philippines, 56

Polesti, bombing raid, 12

Porter, Blackie, 178

Potsdam Conference, 142

PT-17, Stearman PT-17, 20, 22, 23

Quadrant Conference (Aug. '43), 14

Quinn, Chick Marrs (*The Aluminum Trail*), 95

Radio compass (ADF, Automatic Direction Finder), 75-76, 123

Randolph Field, San Antonio, TX, 46-47

Richthofen, Manfred von, 9

Rommel, Field Marshal, Erwin, 83

Roosevelt, Franklin Delano, 2-5, 14, 17, 19, 32-35, 43, 45, 56, 58, 78, 117, 142

Roxboro, N.C. 9, 38, 46

Santsung range, 106

Sevareid, Eric, 6, 178

Sextant, codename for Cairo Conference, 14

Shanghai, China, 148, 174

Shaw Field (Sumter, SC), 47-8

Sinclair, Frank, 19

Soesterberg Air Base, Holland, 180

Solomon Islands, 3

Soong Tze-Ven, 15, 33, 117

Spencer, Otha, 76, 81

St. Elmo's fire, 121

Stalin, Joseph, 14, 56, 142

Stilwell, Gen. Joseph Warren, 17-

19, 31-35, 58, 78, 116-118
Stratemeyer, General George E., 85, 140, 141
Sun Yat-sen, 15

T-50, Cessna "Bamboo Bomber," 46
Tennessee Valley Authority, 43
Tenth (Army) Air Force, 4, 14, 89, 95
Thomas, Anna Rossie Pridmore, 41-42, 45, 67, 69
Thomas, George Washington, 39-41, 44, 46
Thomas, Neva Rae Hudson, 50-52, 61-63, 67-70, 101, 128, 182
Thomas, William Marcus, 40
Thunder out of China (T. White, A. Jacoby), 31-32, 159, 177
Tibbets, Col. Paul, 145
Tokyo Rose, 94, 127, 128, 138
Trident Conference (May '43), 14, 34
Truman, President Harry S., 78, 142, 146, 163, 174
Tunner, General William H., xxv, 22, 59-61, 84-89, 94, 97-99, 101-102, 105, 108, 113, 118, 119, 140-144, 146, 150, 151, 154, 159, 160, 161
turbo-charger, 104

UC-78 Bobcat (Cessna T-50), 46, 47
United Nations Conference, 78
USS Battleship Missouri, 146

Valdosta, Georgia, 35
vertigo, 103

Wake Forest College, 1, 45
Wake Island, 3
Ward, Deming, 2nd Lt., 1- 2, 5, 11, 48, 95-96, 103, 127
Wedemeyer, Gen. Albert C., 35, 58-59, 80, 140-141, 148, 162-164
White, Theodore, 31, 34-35, 142, 159, 177
Willkie, Wendell, 116
Wright, Charles and Orville, 9

Yalta Conference, 56